Current Topics in Microbiology

238 and Immunology

Springer
Berlin
Heidelberg
New York
Barcelona
Hong Kong
London
Milan
Paris
Singapore
Tokyo

Redirection of Th1 and Th2 Responses

Edited by R.L. Coffman and S. Romagnani

With 6 Figures and 10 Tables

 Springer

ROBERT L. COFFMAN, Ph.D.

DNAX Research Institute of
Molecular and Cellular Biology, Inc.
901 California Avenue
Palo Alto, CA 94304-1104
USA

Professor SERGIO ROMAGNANI, M.D.

Head, Istituto di Clinica Medica III
Cattedra di Immunologia Clinica
e Allergologia
Servizio di Immunoallergologia
Viale Morgagni, 85
I-50134 Firenze
Italy

Cover Illustration: Pathogenic role of allergen-specific Th2 responses in allergic inflammation and their regulatory mechanisms.

Cover Design: Design & Production GmbH, Heidelberg

ISSN 0070-217X

ISBN 3-540-65048-2 Springer-Verlag Berlin Heidelberg New York

© Springer-Verlag Berlin Heidelberg 1999
Library of Congress Catalog Card Number 15-12910
Printed in Germany

The use of general descriptive names, registered names, trademarks, etc. in this publication does not imply, even in the absence of a specific statement, that such names are exempt from the relevant protective laws and regulations and therefore free for general use.

Product liability: The publishers cannot guarantee the accuracy of any information about dosage and application contained in this book. In every individual case the user must check such information by consulting other relevant literature.

Typesetting: Scientific Publishing Services (P) Ltd, Madras

Production Editor: Angélique Geouta

SPIN: 10566820   27/3020 – 5 4 3 2 1 0 – Printed on acid-free paper

Preface

It has been 12 years since the first proposal was made to subdivide mouse CD4$^+$ T cell clones into Th1 and Th2 subsets, based on their differences in cytokine production, and 7 years since the first clear demonstration of a similar dichotomy among human T cell clones. In the ensuing period, it has been realized that inappropriate development of Th1 or Th2 responses are important features of many immunological and infectious diseases. Perhaps the first group of diseases to be understood in terms of preferential Th subset activation were allergic diseases (see PARRONCHI et al., this volume). Several of the major, coordinately regulated, features of allergy, including IgE, eosinophilia and mastocytosis, were found to be stimulated by the Th2-specific cytokines IL-4 and IL-5 and inhibited by the Th1 cytokine, IFN-. This suggested that the presence and severity of allergic responses reflected the relative numbers of Th1 and Th2 cells specific for the offending allergen. Similarly, the very different consequences of protective Th1 and nonprotective Th2 responses to a number of intracellular pathogens have been recognized for some time (see TRINCHIERI and SCOTT, and COFFMAN et al., this volume). More recently, it has become recognized that many T cell-mediated autoimmune diseases are mediated by autoreactive Th1 cells and that modest self-reactive Th2 responses can be harmless and even protective (see PEARSON and McDEVITT, this volume). Although it becoming clear that not all of the heterogeneity of function or cytokine production observed in man or animals can be explained by only two discrete Th subsets (see GARCIA and WEINER, this volume), Th1 and Th2 cells remain the two most thoroughly studied and readily defined subsets of CD4$^+$ T cells..

Recognition of the contrasts between Th1 and Th2 responses in many disease states has inevitably led to consideration of the therapeutic possibilities of manipulating the Th1/Th2 balance in chronic diseases and to development of vaccination strategies that produce protective responses of the appropriate subset. Essential to these undertakings is an understanding of the basic mechanisms involved in T cell development. In this volume, the articles by MURPHY et al., COFFMAN et al., and TRINCHIERI and SCOTT review the key events in Th1 and Th2 differentiation and discuss possible mechanisms by which this differentiation process can be controlled and directed. The chapters contributed by PARRONCHI et al. and PEARSON and McDEVITT review human and animal model studies that suggest strategies for T cell-based approaches for the treatment of allergy and autoimmunity. The chapter by GARCIA and WEINER discusses a very different approach to inhibition of specific Th responses, one that exploits a novel subset of regulatory CD4$^+$ T cells that can act as inhibitors of both Th1 and Th2 cells. It is our hope that these papers will help readers navigate through an admittedly complex field and will suggest ways to expand the therapeutic possibilities suggested herein.

SERGIO ROMAGNANI
ROBERT L. COFFMAN

List of Contents

R.L. COFFMAN, S. MOCCI, and A. O'GARRA
The Stability and Reversibility of Th1
and Th2 Populations   1

K.M. MURPHY, W. OUYANG, S.J. SZABO, N.G. JACOBSON,
M.L. GULER, J.D. GORHAM, U. GUBLER, and
T.L. MURPHY
T Helper Differentiation Proceeds Through
Stat1-Dependent, Stat4-Dependent
and Stat4-Independent Phases   13

P. PARRONCHI, E. MAGGI, and S. ROMAGNANI
Redirecting Th2 Responses in Allergy   27

G. TRINCHIERI and P. SCOTT
Interleukin-12: Basic Principles and Clinical Applications .   57

C.I. PEARSON and H.O. McDEVITT
Redirecting Th1 and Th2 Responses
in Autoimmune Disease   79

G. GARCIA and H.L. WEINER
Manipulation of Th Responses by Oral Tolerance 123

Subject Index 147

List of Contributors

(Their addresses can be found at the beginning of their respective chapters.)

COFFMAN, R.L.   1

GARCIA, G.   123

GORHAM, J.D.   13

GUBLER, U.   13

GULER, M.L.   13

JACOBSON, N.G.   13

MAGGI, E.   27

McDEVITT, H.O.   79

MOCCI, S.   1

MURPHY, K.M.   13

MURPHY, T.L.   13

O'GARRA, A.   1

OUYANG, W.   13

PARRONCHI, P.   27

PEARSON, C.I.   79

ROMAGNANI, S.   27

SCOTT, P.   57

SZABO, S.J.   13

TRINCHIERI, G.   57

WEINER, H.L.   123

The Stability and Reversibility of Th1 and Th2 Populations

R.L. Coffman, S. Mocci, and A. O'Garra

1	Introduction	1
2	Stability of Th1 and Th2 cells Generated In Vitro	3
2.1	Reductionist Approach for the Study of Th1 and Th2 Development	3
2.2	The Cytokines Directing the Development of Th1 and Th2 Cells in Reductionist Systems	4
2.3	Can the Phenotype of Th1 or Th2 Cells Generated In Vitro be Reversed?	4
3	The Reversibility of Th Populations Polarized In Vivo	5
3.1	Stability of Th1 and Th2 Populations from *L. major*-Infected Mice	5
3.2	Reversibility in Adoptive Transfer Experiments	6
3.3	Reversibility of Human Th1 and Th2 Populations In Vitro	6
3.4	In Vitro Studies Of Mouse *L. major*-Specific Th1 and Th2 Populations	7
4	The Mechanism of In Vitro Th1 To Th2 Switching in Highly Polarized *Leishmania major*-Specific T Cell Populations	8
References		10

1 Introduction

It has become widely accepted in the past decade that Th1 and Th2 cells represent alternate states of function and gene expression of CD4[1] T cells (Mosmann and Coffman 1989; Abbas et al. 1996). A question central to understanding the relevance of these subsets is whether they are products of an irreversible differentiation process or whether Th1 and Th2 cytokine patterns can be interchanged in an ordered or a regulated manner. Many disease states can be attributed to the activity of one specific Th subset, such as Th1-mediated autoimmune diseases or Th2-mediated allergic diseases and this implies that the ability to alter or reverse Th differentiation is a potential strategy for the treatment of such diseases.

The earliest evidence that Th1 and Th2 cytokine patterns were stable in vitro came from the behavior of mouse Th clones in vitro (Mosmann et al. 1986; Mosmann and Coffman 1987). Many laboratories recognized that cytokine production patterns of most Th clones remained stable after repeated passage, and, in

Department of Immunobiology, DNAX Research Institute of Molecular and Cellular Biology, 901 California Avenue, Palo Alto, CA 94304, USA any event did not change from Th1 to Th2 or vice-versa (MURPHY et al. 1996; BUCY et al. 1994). Although there is little published record of this, we are aware of a number of laboratories that made concerted attempts to change Th1 clones to Th2 or Th2 clones to Th1 with no success.

It is also possible to interpret a great many in vivo observations as providing evidence for the fundamental stability of T cell responses that are dominantly Th1 or Th2. The general experience with secondary responses suggests that memory T cells "remember" both the specificity and the functional phenotype of the primary response (SWAIN 1994; DUTTON et al. 1998). In a few cases, long-term stability of a dominant cytokine pattern has been demonstrated directly (SWAIN 1994). Thus, experimental animals immunized to produce prominent DTH or IgE responses tend to make enhanced responses of the same type upon rechallenge with the antigen. This stability is reflected in human T cell responses as well, with both IgE responses to seasonal allergens and protective Th1-mediated immunity to intracellular pathogens being clear examples of responses that remain stable in individuals for many years or decades.

A substantial number of studies have clearly demonstrated that in many allergic, infectious and autoimmune pathologies, the different disease outcomes correlate with the development of distinct CD4+ T cell subsets (MOSMANN and COFFMAN 1989; SHER and COFFMAN 1992; ABBAS et al. 1996). In particular, experimental infection of mice with the protozoan L. major has provided a useful model for studying both the differentiation of Th1 and Th2 responses as well as the in vivo consequences of these different responses. Thus, the healing response represents the development of IFN-γ-producing Th1 cells, whereas the progressive non-healing response is dominated by IL-4-producing Th2 cells. In such disease, the challenge is to be able to manipulate these populations in situ by changing the phenotype of the dominant Th cells (REINER and LOCKSLEY 1995).

Previous studies have shown that the polarization of Th1 and Th2 responses occurs early in L. major infection and can be easily changed by treatment with anti-IFN-γ and anti-IL-4 Abs, respectively. To be effective these treatments must be given before a strong polarized response is established (CHATELAIN et al. 1992; BELOSEVIC et al. 1989; SYPEK et al. 1993). This suggests, but does not prove, that once a polarized Th1 or Th2 response has been established, the cytokine production pattern becomes relatively stable and resistant to regulatory influences. Alternatively, the failure of anti-IL-4 or anti-IFN-γ treatment to alter an established response could simply reflect an inability of antibodies to completely block cytokine action, or a secondary effect of advanced L. major infection, such as increasing parasite burden and disruption of normal lymphoid architecture. A key question remains as to whether cells that have apparently differentiated into polarized Th1 or Th2 subsets can be influenced to change their cytokine production pattern or even be converted to the opposite cytokine phenotype by using cytokine agonists or antagonists.

Although dominant Th1 and Th2 responses tend to remain stable over time, with either chronic or repeated antigenic stimulation, there are some well-characterized examples of both spontaneous and induced Th reversals in man and experimental animals. In human leprosy, acute episodes termed reactional states occur, in which patients appear to switch from Th2-dominant lepromatous leprosy to Th1-dominant tuberculoid leprosy or vice-versa (COOPER et al. 1989; MODLIN et al. 1986). Such reversals can be spontaneous, but occur more frequently in response to therapeutic treatments. Treatment of lepromatous leprosy patients with IFN-γ, for example, causes a significant frequency of reversals to a more Th1-dominant disease pattern (SAMPAIO et al. 1992). Similarly, successful treatment of visceral leishmaniasis patients leads to a significant increase in parasite-specific Th1 responses and a decrease in Th2-like responses, especially IL-10 (CARVALHO et al. 1985; KARP et al. 1993). In the murine *L. major* infection model described above, mice with a chronic, nonhealing Th2 response can be converted to a dominant Th1 response by treatment with the combination of Il-12 and an anti-leishmanial drug, but not by either therapy alone (NABORS et al. 1995). In an example with a protein antigen, mice primed for a Th2 response to ovalbumin can be converted to a dominant Th1 ovalbumin response by injections of high doses of aggregated ovalbumin (HAYGLASS and STEFURA 1991a,b). This treatment results in an IFN-γ dependent reduction of the IgE response to ovalbumin initially induced by the Th2 cells.

2 Stability of Th1 and Th2 cells Generated In Vitro

2.1 Reductionist Approach for the Study of Th1 and Th2 Development

A number of groups have utilized a reductionist approach to further understand the mechanisms dictating Th1 and Th2 cell development. Initial studies utilized mitogens or anti-T-cell receptor (TCR) antibodies to stimulate CD4$^+$ T cells derived from healthy individuals or non-immunized mice, and tested cytokines and accessory signals to determine their role in directing subset development (SWAIN et al. 1990; LE GROS et al. 1990; MANETTI et al. 1993; MAGGI et al. 1992). The advantage of this is that the role of discrete factors in phenotype development could be studied in the absence of antigen presenting cells (APC), verifying that any effect observed resulted from direct action on the T cell. However, these polyclonal stimulators could also stimulate endogenous cytokine production by memory CD4$^+$ T cells, unless care was taken to use highly purified naïve T cells (GOLLOB and COFFMAN 1994). In order to study Th development in a more physiological, antigen-specific system, several groups have used an antigen-specific $\alpha\beta$-TCR transgenic mouse as a source of naïve CD4$^+$ T cells with a defined antigen specificity (HSIEH et al. 1992; HSIEH et al. 1993; SEDER et al. 1992; SEDER et al. 1993). In this system, naïve T cells are stimulated with antigen processed and presented by antigen-presenting cells (APC). The interpretation of such experiments can be more difficult, however, as the APC can be a source of endogenous cytokines as well as a target of any added cytokines.

2.2 The Cytokines Directing the Development of Th1 and Th2 Cells in Reductionist Systems

Both IL-12 and IFN-γ have been shown to be direct inducers of the development of Th1 cells producing high levels of IFN-γ (SWAIN et al. 1990; MAGGI et al. 1992; SEDER et al. 1993; HSIEH et al. 1993; GOLLOB and COFFMAN 1994). The role of IL-12 in Th1 cell development has been demonstrated both by the addition of exogenous sources of IL-12 during stimulation of naïve Th cells with antigen or mitogen (HSIEH et al. 1993; MACATONIA et al. 1993; SEDER et al. 1993; MANETTI et al. 1993; SORNASSE et al. 1996) and by the addition of highly activated macrophages or dendritic cells producing IL-12 (HSIEH et al. 1993; MACATONIA et al. 1995). In addition to directing the development of Th1 cells from naïve CD4$^+$ T cells, IL-12 can also potentiate the production of IFN-γ from already differentiated Th1 populations and clones (MURPHY et al. 1996; ROBINSON et al. 1997). This suggests that even when Th1 cells are fully differentiated, they require stimulation with IL-12 for optimum levels of IFN-γ production. Using the above reductionist approaches, it is also clear that IL-4 during primary culture of antigen or mitogen-stimulated CD4$^+$ T cells results in the development of Th2 cells, producing high levels of IL-4 and undetectable levels of IFN-γ upon subsequent restimulation (SWAIN et al. 1990; LE GROS et al. 1990; HSIEH et al. 1992; SEDER et al. 1992). As with IFN-γ and IL-12, the IL-4 can be added to cultures or supplied endogenously by other cell types (GOLLOB and COFFMAN 1994).

2.3 Can the Phenotype of Th1 or Th2 Cells Generated In Vitro be Reversed?

The simple culture systems described above have also been used to study the stability or reversibility of newly differentiated Th1 and Th2 populations. Th1 cells derived from naïve TCR-transgenic CD4$^+$ T cells by one round of stimulation with antigen and APC plus IL-12 can be readily converted to Th2 populations by a second round of stimulation with antigen and APC plus IL-4 (SZABO et al. 1995; PEREZ et al. 1995). Despite the fact that the T cells are clonal with respect to their antigen receptor, these cultures are actually heterogeneous populations of Th cells with respect to their cytokine production (MURPHY et al. 1996; OPENSHAW et al. 1995). For example, polarized Th1 populations derived from naïve TCR-transgenic CD4$^+$ T cells after one round of antigenic stimulation in the presence of IL-12 secrete only Th1 cytokines into the culture supernatants. However, analyses for cytokine production at the single cell level show that these populations are indeed heterogeneous, with a majority of the cells producing neither IFN-γ or IL-4 (MURPHY et al. 1996) Thus, the conversion of a such Th1 to a Th2 population may actually represent development of Th2 cells from undifferentiated precursor T cells or expansion of a small number of contaminating, differentiated Th2 cells. However, the generation of Th2 cells in these secondary cultures is accompanied by the disappearance of the IFN-γ producing cells that were generated in the primary cultures, suggesting that the process is more complex than just the generation of new Th2 cells from precursors. In contrast, long-term Th1 clones or more homogeneous Th1 populations derived from repeated stimulation with antigen under Th1-polarizing conditions cannot be converted to the Th2 phenotype by restimulation in the presence of IL-4. Repeated stimulation of these irreversibly committed Th1 populations with antigen and APC in the presence of IL-4 could significantly reduce the levels of IFN-γ produced by these Th1 cells, but did not give rise to IL-4 producing cells.

In contrast to the ease with which the phenotype of primary Th1 populations can be reversed, early Th2 cells are relatively resistant in some cases to phenotype reversal (PEREZ et al. 1995; PEREZ et al. 1995). The inability to convert such Th2 populations – derived from one round of antigenic stimulation of TCR-transgenic T cells in IL-4 – to Th1 cells by restimulating with antigen in the presence of IL-12 has been shown to reflect a rapid loss of responsiveness of developing Th2 cells to IL-12 (SZABO et al. 1995; SZABO et al. 1997; ROGGE et al. 1997). In our laboratory, primary populations of Th2 cells could give rise to Th1 cells upon antigenic restimulation in IL-12. In these experiments, however, this appeared to be the result of development from uncommitted precursor T cells or the expansion of rare, previously committed Th1 cells, as the IL-4 producing population remained numerically identical after secondary culture with IL-12 (MURPHY et al. 1996). However, Th2 populations and clones stimulated multiple times could not give rise to IFN-γ producing cells under the same conditions. These data contrast with studies in human systems where short-term Th2 cells derived from mitogen-stimulated cord blood cells can still be induced to produce IFN-γ if restimulated in the presence of IL-12 (SORNASSE et al. 1996). This may reflect basic differences in the extent of differentiation of Th populations in mouse and human culture systems or may simply reflect differences in the regulation of IL-12 receptor expression (SZABO et al. 1995, 1997; ROGGE et al. 1997). Taken together, these studies suggest that it may be more difficult to change or reduce an IL-4-producing Th2 population than a IFN-γ-producing Th1 population. The differences in reversibility between populations generated by one versus multiple rounds of antigen stimulation and cytokine exposure suggest one of two possibilities. Either repeated stimulations are required for individual Th cells (or their descendants) to become irreversibly committed to one cytokine pattern or there is a progressive loss in populations of na-like cells capable of being induced to become either Th1 or Th2 cells.

3 The Reversibility of Th Populations Polarized In Vivo

3.1 Stability of Th1 and Th2 Populations from *L. major*-Infected Mice

As discussed earlier in this article, the Th1 and Th2 responses of mice infected with *L. major* are particularly well-characterized examples of responses that are very resistant to change in vivo. Within 2–3 weeks of infection, both healing Th1 and nonhealing Th2 responses were resistant to reversal or significant alteration by treatment with anti-IL-4, anti-IFN-γ or IL-12, all treatments that were able to reverse the Th differentiation pattern within the first week of infection. Only the combination of IL-12 and a parasiticidal drug has been shown capable of effecting a substantial change from a Th2 to a Th1 response to *L. major* (NABORS et al. 1995).

3.2 Reversibility in Adoptive Transfer Experiments

However, we have made the rather surprising observation that Th populations isolated from lesion-draining lymph nodes (LN) of *L. major*-infected mice 4 weeks or more after infection could be readily changed from Th1 to Th2 or from Th2 to Th1 when transferred to *L. major*-infected C.B-17 *scid* mice (POWRIE et al. 1994a; SEDER et al. 1994). This switch in Th responses could be induced by altering the relative levels of IFN-γ and IL-4 with neutralizing mAbs to one or the other. Thus, transfer of 10^6 LN cells from a nonhealing (Th2) BALB/c mouse into T and B cell-deficient C.B-17 *scid* mice would transfer, as expected, a nonhealing Th2 response to *L. major* infection. If recipient mice were given a single injection of anti-IL-4 at the time of transfer, however, the donor cells conferred a stable, dominant Th1 response that conferred long-lived resistance to *L. major* infection. This demon-strated that cytokine profiles of parasite-specific Th1 and Th2 populations were capable of extensive modification under certain conditions, long after they were effectively irreversible in the primary infected host. In a similar study, an uncloned, *L. major*-specific Th2 cell line could give rise to a protective Th1 response when transferred to C.B-17 *scid* mice treated with anti-IL-4 mAb (HOLADAY et al. 1991). Thus, adoptive transfer experiments reveal a potential for reversal of Th1 and Th2 populations that is difficult to accomplish in situ.

3.3 Reversibility of Human Th1 and Th2 Populations In Vitro

A similar potential for reversal of cytokine patterns in human has been demon-strated in in vitro cultures of Th1 populations from vaccinated individuals and Th2 populations from allergic patients (PARRONCHI et al. 1992). Culture of peripheral blood T cells for two weeks with antigen, IL-2 and IL-4 was able to cause a marked Th1 to Th2 shift in purified protein-specific Th cells and, likewise, culture of Th cells from allergic individuals with allergen, IL-2 and either IFN-α or IFN-γ was able to convert the responses from Th2 to Th1 on restimulation. This important study was the first demonstration that reversal of highly polarized populations could be accomplished and studied in vitro, and it implied that therapeutic reversal of Th responses, even after years of repeated allergen exposure, was at least the-oretically possible.

3.4 In Vitro Studies Of Mouse *L. major*-Specific Th1 and Th2 Populations

To further understand the factors and mechanisms that influence cytokine production by highly differentiated Th populations, we have established an in vitro culture system by using Th populations from mice infected with *L. major* for at least 4 weeks (Mocci and Coffman 1995). By this time after infection with *L. major*, the T cell responses were highly polarized and could not be significantly modified in vivo by treatment with IL-4, IFN-γ, or the neutralizing Abs to these cytokines (Sadick et al. 1990, 1991; Chatelain et al. 1992; Belosevic et al. 1989; Coffman et al. 1991). Purified CD4 ¹ T cells were cultured with crude *L. major* antigens (LmAg), IL-2 and T cell-depleted splenocytes as APC. This was essentially the same technique used by Parronchi et al. (1992) for human T cells, with the exception that cells were cultured no longer than one week before restimulation. Various combinations of cytokines and/or anti-cytokine antibodies were added at the beginning of the cultures to attempt to modify the cytokine profile and, at the end of the culture period, cells were harvested, restimulated with fresh antigen and APC and the supernatants collected after 60–72h for cytokine analysis.

The most decisive change observed in these experiments was the conversion of a polarized Th1 population to a Th2 cytokine pattern by culture with LmAg and IL-4 for 7 days. After examining the effects of many other cytokines, anti-cytokine antibodies and combinations, it became clear that IL-4 was both necessary and sufficient to induce the switch from Th1 to Th2 response. No other Th2-specific cytokines – including IL-13, which shares many activities of IL-4 (Doherty et al. 1993; de Waal Malefyt et al. 1993), and IL-10, which can inhibit Th1 cytokine production (Fiorentino et al. 1991a,b) – were able to cause any significant change in the initial Th1 population. Most striking was the fact that, although high levels of IFN-γ were initially produced by Th1 cells responding to antigen in the first culture period, neutralization of this cytokine did not result in a Th1 to Th2 shift nor did it enhance the ability of IL-4 to induce this switch. In cultures of naïve Th cells, IL-4 mediated Th2 differentiation is dominant over IFN-γ mediated Th1 differentiation, when both cytokines are present in high concentration (Seder et al. 1992), and this appears be the case for the development of a Th2 response from a primed Th1 population as well.

The switch from dominant Th1 to Th2 cytokine pattern induced by IL-4 in vitro represents the generation of a stable Th2 population, rather than just a transient induction of Th2 and inhibition of Th1 cytokines. This was demonstrated by the ability of these cells to remain a dominant Th2 response when adoptively transferred to *L. major*-infected C.B-17 *scid* mice (Mocci and Coffman 1995).

IL-4 induction of the Th1 to Th2 switch in the response to *L. major* is quite rapid (Mocci and Coffman 1995). Reduced IFN-γ production and significant induction of IL-4, IL-5 and IL-10 were observed in cultures restimulated after only 5 days of primary culture with IL-4. The critical time for exposure to IL-4 was defined in a series of experiments in which IL-4 was either added at various times after the initiation of the cultures or was added at the beginning of the culture and neutralized with anti-IL-4 Abs at various later times. Both types of analysis showed that IL-4 exposure during the first 24 h of culture was not able to induce differentiation to a Th2 population. Rather, the critical time for IL-4 exposure was from day 1 to 4 of culture and, for maximum conversion, IL-4 needed to be present for this entire period. These data suggest that IL-4 may play a prominent role during the proliferative phase of T cell activation, rather than during the initial trigger of the T cell receptor.

The switch, in vitro, from an *L. major*-specific Th2 to a Th1 population was somewhat less efficient (Mocci and Coffman 1995), despite the fact that the non-healing response in BALB/c mice contains a small Th1 component that is actively suppressed in vivo by IL-4 and IL-10 (Powrie et al. 1994b). Both IFN-γ and IL-12 induced measurable IFN-γ production upon restimulation, but substantially more IFN-γ production was obtained when anti-IL-4 antibody was also added to the cultures. Even with all three components added, however, significant levels of IL-4 and IL-10 were produced upon recall after only 1 week of culture. After two or three weekly cycles of stimulation with anti-IL-4, IFN-γ, IL-12 and antigen, however, only Th1-specific cytokines were made in significant amounts by these populations (S.M. and R.L.C., unpublished).

Taken together, these data indicate that the maintenance of secondary Th1 and Th2 responses, like the development of primary ones, was dependent on IFN-γ and IL-4, respectively, and that these polarized CD4+ T cell specific for *L. major* retained the capacity to develop quite different cytokine profiles under appropriate conditions. The possibility of altering established Th1 and Th2 responses at the population level has considerable therapeutic implications. In many clinical situations patients present with already established Th response, so the challenge is to be able to manipulate these populations in situ.

4 The Mechanism of In Vitro Th1 To Th2 Switching in Highly Polarized *Leishmania major*-Specific T Cell Populations

It is important to stress that, in all of the studies discussed so far, the Th subset shifts were induced in populations of cells, not in individual T cells or clones. Even in the experiments of Holaday et al., the *L. major*-specific Th2 cell line that could be switched to Th1 on adoptive transfer and treatment with anti-IL-4 was multiclonal, and clonal lines derived from it were not able to be switched to Th1 (Holaday et al. 1991). Two quite distinct possibilities exist for the cellular events that take place when such a switch occurs. Using as an example the Th2 switch induced by culture of Th1 cells with IL-4 and antigen for one week, it is possible that the Th2 cells at the end of the culture period were direct descendants of cells that produced only Th1 cytokines before the addition of IL-4. But it is equally possible that these Th2 cells were derived from a very small number of Th2 cells in the starting population or from antigen-specific T cells that had not differentiated fully into either Th1 or Th2 cells.

The strategy we have used to determine the mechanism of the Th1 to Th2 switch after culture with IL-4 is to determine whether a direct precursor/product relationship exists between Th1 cells at the beginning of the culture and the Th2 cells that are present at the end of the culture period (Mocci and Coffman 1997). For this, CD4$^+$ Th1 cells from the DLN were further separated by the level of L-selectin expression, using the Mel-14 mAb (Gallatin et al. 1983). The CD4$^+$ Mel-14^{lo} subset has been shown to include nearly all of the cells that have previously been activated by antigen (Bradley et al. 1992), whereas the Mel-14^{hi} subset includes the great majority of na cells that have not encountered Ag or acquired the ability to make cytokines other than IL-2 when stimulated (Bradley et al. 1992; Croft et al. 1992; Mackay et al. 1990) Mel-14^{lo} and Mel-14^{hi} CD4$^+$ T cells were then cultured with LmAg, splenic APC and IL-2, in the presence or absence of IL-4. After 7 days, the phenotype of each cell subset was tested by restimulating the cells with fresh APC and LmAg and measuring the resulting production of IL-4 and IFN-γ. Although the unseparated CD4$^+$ T cell population switched essentially completely to a Th2 response, the Mel-14^{lo} cells retained their Th1 phenotype whether cultured with or without IL-4 (Mocci and Coffman 1997). The Mel-14^{hi} fraction, which contained neither Th1 or Th2 cells at the initiation of the cultures, generated cells with a Th2 cytokine secretion pattern when cultured with IL-4. These results demonstrate that the *L. major*-specific Th2 cells induced from a Th1 polarized population could not be derived directly from cells with a Th1 pattern of cytokine expression, as they were derived instead from a subpopulation of CD4$^+$ cells that contained neither Th1 or Th2 cells. The same conclusion was drawn from similar experiments converting fractions of Th2 populations to a Th1 pattern (S.M. and R.L.C., unpublished).

Thus, the highly polarized *L. major*-specific population within the DLN of a healing mouse appeared to be heterogeneous, containing both irreversibly differentiated Mel-14^{lo} Th1 cells and Mel-14^{hi} cells that were undifferentiated with regard to both cytokine production and cell surface marker expression. Moreover, the undifferentiated Mel-14^{hi} Th precursor population could be induced to develop into either Th1 or Th2 cells, as shown by the generation of a strong Th1 population after 1 week of culture in the presence of IL-12 and anti-IL-4-neutralizing Ab, suggesting that the cells were not yet committed to either Th phenotype (Mocci and Coffman 1997). Whether the Mel-14^{hi} CD4$^+$ T cells are truly naïve or whether they had previously encountered antigen, yet maintained a naïve cell surface phenotype is not yet known with certainty. We have observed that *L. major*-specific Th2 cells could only be generated from "naïve like" Mel-14^{hi} CD4$^+$ T cells from *L. major*-infected mice, not from non-infected mice. This suggests that the pool of *L. major*-specific precursors was substantially expanded within the CD4$^+$ Mel-14^{hi} cells of infected mice.

These analyses of the stability and reversibility of Th populations suggest that the encounter of a naïve CD4$^+$ T cell with presented antigen and appropriate cytokines has two possible outcomes. The first is the well-studied differentiation into a cell restricted to produce either the Th1 or Th2 cytokine pattern and unable to be subsequently induced to produce the opposite pattern of cytokines. Given the stability of long term Th clones, it appears that this differentiation state is inherited by all descendants of this cell, in other words, the clone of which it is the founder. We suggest that the second possibility is the regeneration, accompanied by proliferation, of cells functionally and phenotypically indistinguishable from the original naïve cell and able to undergo either Th1 or Th2 differentiation during a subsequent encounter with antigen. Thus, although fully differentiated Th1 and Th2 cells in vivo may have a fixed cytokine production pattern, apparently polarized populations can retain the ability to develop responses of the opposite phenotype. It is therefore possible that in chronic diseases characterized by a dominant Th1 or Th2 phenotype, there is still a population of Th precursor cells that has been primed and expanded by the antigen, but has retained a naïve undifferentiated phenotype. Further work is needed to determine how best to deliver cytokines and antigens to induce expansion and appropriate differentiation of these precursors for the treatment of chronic, inappropriate Th responses.

Acknowledgement. The DNAX Research Institute is supported by the Schering-Plough Corporation.

References

Abbas AK, Murphy KM, Sher A (1996) Functional diversity of helper T lymphocytes. Nature 383:787 793
Belosevic M, Finbloom DS, Van Der Meide PH, Slayter MV, Nacy CA (1989) Administration of monoclonal anti-IFN-gamma antibodies in vivo abrogates natural resistance of C3H/HeN mice to infection with Leishmania major. J Immunol 143:266 274
Bradley LM, Atkins GG, Swain SL (1992) Long-term CD4+ memory T cells from the spleen lack MEL-14, the lymph node homing receptor. J Immunol 148:324 331
Bucy RP, Panoskaltsis-Mortari A, Huang GQ, Li J, Karr L, Ross M, Russell JH, Murphy KM, Weaver CT (1994) Heterogeneity of single cell cytokine gene expression in clonal T cell populations. J Exp Med 180:1251 1262
Carvalho EM, Badaro R, Reed SG, Jones TC, Johnson WD Jr (1985) Absence of gamma interferon and interleukin 2 production during active visceral leishmaniasis. J Clin Invest 76:2066 2069
Chatelain R, Varkila K, Coffman RL (1992) IL-4 induces a Th2 response in Leishmania-major-infected mice. J Immunol 148:1182 1187
Coffman RL, Varkila K, Scott P, Chatelain R (1991) The role of cytokines in the differentiation of CD4+ T cell subsets in vivo. Immunol Rev 123:189 207
Cooper CL, Mueller C, Sinchaisri TA, Pirmez C, Chan J, Kaplan G, Young SM, Weissman IL, Bloom BR, Rea TH, et al (1989) Analysis of naturally occurring delayed-type hypersensitivity reactions in leprosy by in situ hybridization. J Exp Med 169:1565 1581
Croft M, Duncan DD, Swain SL (1992) Response of naive antigen-specific CD4+ T cells in vitro: characteristics and antigen-presenting cell requirements. J Exp Med 176:1431 1437
de Waal Malefyt R, Figdor CG, Huijbens R, Mohan-Peterson S, Bennett B, Culpepper J, Dang W, Zurawski G, de Vries JE (1993) Effects of IL-13 on phenotype, cytokine production, and cytotoxic function of human monocytes. Comparison with IL-4 and modulation by IFN-gamma or IL-10. J Immunol 151:6370 6381
Doherty TM, Kastelein R, Menon S, Andrade S, Coffman RL (1993) Modulation of murine macrophage function by IL-13. J Immunol 151:7151 7160
Dutton RW, Bradley LM, Swain SL (1998) T cell memory. Annu Rev Immunol 16:201 223

Fiorentino DF, Zlotnik A, Mosmann TR, Howard M, O'Garra A (1991a) IL-10 inhibits cytokine production by activated macrophages. J Immunol 147:3815 3822

Fiorentino DF, Zlotnik A, Vieira P, Mosmann TR, Howard M, Moore KW, O'Garra A (1991b) IL-10 acts on the antigen-presenting cell to inhibit cytokine production by Th1 cells. J Immunol 146:3444 3451

Gallatin WM, Weissman IL, Butcher EC (1983) A cell-surface molecule involved in organ-specific homing of lymphocytes. Nature 304:30 34

Gollob KJ, Coffman RL (1994) A minority subpopulation of CD4 + T cells directs the development of naive CD4 + T cells into IL-4-secreting cells. J Immunol 152:5180 5188

Hayglass KT, Stefura BP (1991a) Anti-interferon-gamma treatment blocks the ability of glutaraldehyde-polymerized allergens to inhibit specific IgE responses. J Exp Med 173:279 285

Hayglass KT, Stefura WP (1991b) Antigen-specific inhibition of ongoing murine IgE responses .2. Inhibition of IgE responses induced by treatment with glutaraldehyde-modified allergens is paralleled by reciprocal increases in IgG2a synthesis. J Immunol 147:2455 2460

Holaday BJ, Sadick MD, Wang ZE, Reiner SL, Heinzel FP, Parslow TG, Locksley RM (1991) Reconstitution of Leishmania immunity in severe combined immunodeficient mice using Th1- and Th2-like cell lines. J Immunol 147:1653 1658

Hsieh CS, Heimberger AB, Gold JS, O'Garra A, Murphy KM (1992) Differential regulation of T helper phenotype development by interleukins 4 and 10 in an alpha beta T-cell-receptor transgenic system. Proc Natl Acad Sci USA 89:6065 6069

Hsieh CS, Macatonia SE, Tripp CS, Wolf SF, O'Garra A, Murphy KM (1993) Development of TH1 CD4 + T cells through IL-12 produced by Listeria-induced macrophages. Science 260:547 549

Karp CL, el-Safi SH, Wynn TA, Satti MM, Kordofani AM, Hashim FA, Hag-Ali M, Neva FA, Nutman TB, Sacks DL (1993) In vivo cytokine profiles in patients with kala-azar. Marked elevation of both interleukin-10 and interferon-gamma [see comments]. J Clin Invest 91:1644 1648

Le Gros G, Ben-Sasson SZ, Seder R, Finkelman FD, Paul WE (1990) Generation of interleukin 4 (IL-4)-producing cells in vivo and in vitro: IL-2 and IL-4 are required for in vitro generation of IL-4-producing cells. J Exp Med 172:921 929

Macatonia SE, Hsieh CS, Murphy KM, O'Garra A (1993) Dendritic cells and macrophages are required for Th1 development of CD4 + T cells from alpha beta TCR transgenic mice: IL-12 substitution for macrophages to stimulate IFN-gamma production is IFN-gamma-dependent. Int Immunol 5:1119 1128

Macatonia SE, Hosken NA, Litton M, Vieira P, Hsieh CS, Culpepper JA, Wysocka M, Trinchieri G, Murphy KM, O'Garra A (1995) Dendritic cells produce IL-12 and direct the development of Th1 cells from naive CD4 + T cells. J Immunol 154:5071 5079

Mackay CR, Marston WL, Dudler L (1990) Naive and memory T cells show distinct pathways of lymphocyte recirculation. J Exp Med 171:801 817

Maggi E, Parronchi P, Manetti R, Simonelli C, Piccinni MP, Rugiu FS, De Carli M, Ricci M, Romagnani S (1992) Reciprocal regulatory effects of IFN-gamma and IL-4 on the in vitro development of human Th1 and Th2 clones. J Immunol 148:2142 2147

Manetti R, Parronchi P, Giudizi MG, Piccinni MP, Maggi E, Trinchieri G, Romagnani S (1993) Natural killer cell stimulatory factor (interleukin 12 [IL-12]) induces T helper type 1 (Th1)-specific immune responses and inhibits the development of IL-4-producing Th cells. J Exp Med 177:1199 1204

Mocci S, Coffman RL (1995) Induction of a Th2 population from a polarized Leishmania-specific Th1 population by in vitro culture with IL-4. J Immunol 154:3779 3787

Mocci S, Coffman RL (1997) The mechanism of in vitro T helper cell type 1 to T helper cell type 2 switching in highly polarized Leishmania major-specific T cell populations. J Immunol 158:1559 1564

Modlin RL, Mehra V, Jordan R, Bloom BR, Rea TH (1986) In situ and in vitro characterization of the cellular immune response in erythema nodosum leprosum. J Immunol 136:883 886

Mosmann TR, Cherwinski H, Bond MW, Giedlin MA, Coffman RL (1986) Two types of murine helper T cell clone. 1. Definition according to profiles of lymphokine activities and secreted proteins. J Immunol 136:2348 2357

Mosmann TR, Coffman RL (1987) Two types of mouse helper T cell clone: implications for immune regulation. Immunol Tod 8:223 227

Mosmann TR, Coffman RL (1989) Th1 and Th2 cells: different patterns of lymphokine secretion lead to different functional properties. Ann Rev Immunol 7:145 173

Murphy E, Shibuya K, Hosken N, Openshaw P, Maino V, Davis K, Murphy K, O'Garra A (1996) Reversibility of T helper 1 and 2 populations is lost after long-term stimulation. J Exp Med 183:901 913

Nabors GS, Afonso LC, Farrell JP, Scott P (1995) Switch from a type 2 to a type 1 T helper cell response and cure of established Leishmania major infection in mice is induced by combined therapy with interleukin 12 and Pentostam. Proc Natl Acad Sci USA 92:3142 3146

Openshaw P, Murphy EE, Hosken NA, Maino V, Davis K, Murphy K, O'Garra A (1995) Heterogeneity of intracellular cytokine synthesis at the single-cell level in polarized T helper 1 and T helper 2 populations. J Exp Med 182:1357 1367

Parronchi P, De Carli M, Manetti R, Simonelli C, Sampognaro S, Piccinni MP, Macchia D, Maggi E, Del Prete G, Romagnani S (1992) IL-4 and IFN (alpha and gamma) exert opposite regulatory effects on the development of cytolytic potential by Th1 or Th2 human T cell clones. J Immunol 149:2977 2983

Perez VL, Lederer JA, Lichtman AH, Abbas AK (1995) Stability of Th1 and Th2 populations. Int Immunol 7:869 875

Powrie F, Coffman RL, Correa-Oliveira R (1994a) Transfer of CD4+ T cells to C.B-17 SCID mice: a model to study Th1 and Th2 cell differentiation and regulation in vivo. Res Immunol 145:347 353

Powrie F, Correa-Oliveira R, Mauze S, Coffman RL (1994b) Regulatory interactions between CD45RB^high and CD45RB^low CD4+ T cells are important for the balance between protective and pathogenic cell-mediated immunity. J Exp Med 179:589 600

Reiner SL, Locksley RM (1995) The regulation of immunity to Leishmania major. Annu Rev Immunol 13:151 177

Robinson D, Shibuya K, Mui A, Zonin F, Murphy E, Sana T, Hartley SB, Menon S, Kastelein R, Bazan F, O'Garra A (1997) IGIF does not drive Th1 development but synergizes with IL-12 for interferon-gamma production and activates IRAK and NFkappaB. Immunity 7:571 581

Rogge L, Barberis-Maino L, Biffi M, Passini N, Presky DH, Gubler U, Sinigaglia F (1997) Selective expression of an interleukin-12 receptor component by human T helper 1 cells. J Exp Med 185:825 831

Sadick MD, Heinzel FP, Holaday BJ, Pu RT, Dawkins RS, Locksley RM (1990) Cure of murine leishmaniasis with anti-interleukin 4 monoclonal antibody. Evidence for a T cell-dependent, interferon gamma-independent mechanism. J Exp Med 171:115 127

Sadick MD, Street N, Mosmann TR, Locksley RM (1991) Cytokine regulation of murine leishmaniasis: interleukin 4 is not sufficient to mediate progressive disease in resistant C57BL/6 mice. Infect Immun 59:4710 4714

Sampaio EP, Moreira AL, Sarno EN, Malta AM, Kaplan G (1992) Prolonged treatment with recombinant interferon gamma induces erythema nodosum leprosum in lepromatous leprosy patients. J Exp Med. 175:1729 1737

Seder RA, Paul WE, Davis MM, Fazekas de St.Groth B (1992) The presence of interleukin 4 during in vitro priming determines the lymphokine-producing potential of CD4+ T cells from T cell receptor transgenic mice. J Exp Med 176:1091 1098

Seder RA, Gazzinelli R, Sher A, Paul WE (1993) Interleukin 12 acts directly on CD4+ T cells to enhance priming for interferon gamma production and diminishes interleukin 4 inhibition of such priming. Proc Natl Acad Sci USA 90:10188 10192

Seder RA, Germain RN, Linsley PS, Paul WE (1994) CD28-mediated costimulation of interleukin 2 (IL-2) production plays a critical role in T cell priming for IL-4 and interferon gamma production. J Exp Med 179:299 304

Sher A, Coffman RL (1992) Regulation of immunity to parasites by T cells and T cell-derived cytokines. Annu Rev Immunol 10:385 409

Sornasse T, Larenas PV, Davis KA, de Vries JE, Yssel H (1996) Differentiation and stability of T helper 1 and 2 cells derived from naive human neonatal CD4+ T cells, analyzed at the single-cell level. J Exp Med 184:473 483

Swain SL, Weinberg AD, English M, Huston G (1990) IL-4 directs the development of Th2-like helper effectors. J Immunol 145:3796 3806

Swain SL (1994) Generation and in vivo persistence of polarized Th1 and Th2 memory cells. Immunity 1:543 552

Sypek JP, Chung CL, Mayor SE, Subramanyam JM, Goldman SJ, Sieburth DS, Wolf SF, Schaub RG (1993) Resolution of cutaneous leishmaniasis: interleukin 12 initiates a protective T helper type 1 immune response. J Exp Med 177:1797 1802

Szabo SJ, Jacobson NG, Dighe AS, Gubler U, Murphy KM (1995) Developmental commitment to the Th2 lineage by extinction of IL-12 signaling. Immunity 2:665 675

Szabo SJ, Dighe AS, Gubler U, Murphy KM (1997) Regulation of the interleukin (IL)-12R beta 2 subunit expression in developing T helper 1 (Th1) and Th2 cells. J Exp Med 185:817 824

T Helper Differentiation Proceeds Through Stat1-Dependent, Stat4-Dependent and Stat4-Independent Phases

K.M. Murphy, W. Ouyang, S.J. Szabo, N.G. Jacobson, M.L. Guler,
J.D. Gorham, U. Gubler, and T.L.Murphy

1 Multiple Parameters Influence Th1/Th2 Development   13

2 Interferon-γ Modifies Early T Cell Responses to Interleukin-12   14

3 The Interleukin-12 Signaling Pathway Is a Point of Regulation for Stabilizing
   Developing T Helper Phenotype Responses   15

4 Roles of Stat1 and Stat4 in Th1/Th2 Development:
   Separation of Distinct Phases of Development   17

5 Effects of Interferon-α on Th1/Th2 Development: Differences Between Human
   and Murine Signaling   18

6 Regulation of Interferon-γ Signaling in Developing Th1 and Th2 Populations   20

7 Genetic Effects on Early Th1/Th2 Development   20

8 Summary   22

References   23

1 Multiple Parameters Influence Th1/Th2 Development

The effort to understand Th1 and Th2 development has included defining the specific signals that determine phenotype fate upon primary T cell activation by antigen. Numerous parameters of T cell activation appear to influence the overall balance of Th1/Th2 phenotype development, including the antigen presenting cells (APCs) used for T cell priming (CHANG et al. 1990), antigen dose (PARISH and LIEW 1972; HOSKEN et al. 1995; MURRAY et al. 1992; CONSTANT et al. 1995), antigen structure or particularly the affinity for the major histocompatibility complex (MHC) and T cell receptor (TCR) (MURRAY et al. 1992; PFEIFFER et al. 1991), levels of costimulation during T cell priming (FREEMAN et al. 1995; LENSCHOW et al. 1995; KUCHROO et al. 1995), genetic background (MURPHY et al. 1994; KUBIN et al. 1994), pathogen-derived materials, and cytokines present in the priming milieu (LE GROS et al. 1990; SWAIN et al. 1990, 1991; SADICK et al. 1990; MAGGI et al. 1992;

Howard Hughes Medical Institute, Department of Pathology/Center for Immunology, Washington University School of Medicine/Box 8118, 660 S. Euclid Avenue, St. Louis, MO 63110, USA

MANETTI et al. 1993; SEDER et al. 1993; SYPEK et al. 1993; CHATELAIN et al. 1992; BELOSEVIC et al. 1989; HSIEH et al. 1995; HOWARD 1986; HEINZEL et al. 1989; SCOTT et al. 1988; LOCKSLEY and SCOTT 1991). While any of these parameters can alter the overall Th1/Th2 developmental balance, some appear to act directly to deliver final Th1/Th2 inducing signals to the T cell, while others appear to act indirectly, for example through modifying APC function, other innate immune cell activity, or the levels of Th1/Th2 inducing cytokines. The cytokines interleukin (IL)-12 and IL-4 act directly on receptors expressed by activated T cells, through specific STAT factors, to deliver direct differentiation-inducing signals (KAPLAN et al. 1996a,b; THIERFELDER et al. 1996). IL-4 activation of Stat6 is necessary for IL-4-induced Th2 phenotype development (LE GROS et al. 1990; SWAIN et al. 1990; MAGGI et al. 1992; CHATELAIN et al. 1992; KAPLAN et al. 1996a; KOPF et al. 1993; KUHN et al. 1991; BETZ and FOX 1990; HOU et al. 1994; QUELLE et al. 1995; SHIMODA et al. 1996; TAKEDA et al. 1996), and IL-12 activation of Stat4 is necessary for Th1 development (KAPLAN et al. 1996b; THIERFELDER et al. 1996; JACOBSON et al. 1995; BACON et al. 1995; MATTNER et al. 1996; MAGRAM et al. 1996; SZABO et al. 1995). While the molecular downstream targets of Stat6 and Stat4 for Th1/Th2 development are currently unknown, at present these two factors are the most proximal known signals controlling phenotype. It is unresolved at present through what mechanisms non-cytokine parameters influence Th1/Th2 balance, although it is likely that some may act by altering the initial levels of IL-4, IL-12 or interferon (IFN)-γ available to T cells during primary activation. Whether partial signaling through the TCR acts directly to induce Th1/Th2 developmental signals is an open issue at present. Changing antigen dose can cause apparent changes in phenotype development in vitro (HOSKEN et al. 1995). However, this effect was lost when IL-4 was neutralized, suggesting that cytokines are dominant in the hierarchy of these parameters.

2 Interferon-γ Modifies Early T Cell Responses to Interleukin-12

We and other investigators had identified IL-12 as a potent inducer of Th1 development in vitro (MANETTI et al. 1993; SEDER et al. 1993; MACATONIA et al. 1993). However, even in our earliest studies, we saw that IFN-γ was a necessary but not sufficient factor for Th1 development in vitro induced by pathogen, such as heat-killed *Listeria monocytogenes* (HKLM). When IFN-γ was neutralized in cultures where Th1 development had been induced by the addition of *Listeria*, we noted a significant reduction in the overall level of Th1 development. Thus neutralizing IFN-γ appeared to be blocking HKLM-induced Th1 development.

Neutralizing IFN-γ could act at two levels. First, it could act on the APCs to diminish the induction of IL-12 by the *Listeria*. Second, it could act on the T cell, perhaps by preventing a necessary action on the T cell for Th1 development. In our study on the regulation of IL-12 p40 promoter (MURPHY et al. 1995), we had noted that IFN-γ augments IL-12 production stimulated by HKLM. These effects of IFN-γ could be mediated even at the level of a reporter construct acting within the first 200 base pairs of the p40 promoter. While the most significant *cis* element residing within this region is the NFκ-B site, induced by lipopolysaccharide (LPS) or lipoteichoic acid (LTA; CLEVELAND et al. 1996), the specific and relevant site for mediating IFN-γ augmentation of p40 has not been identified. Within this region, there is no perfect consensus for Stat1 or for one that can be easily demonstrated to bind with high affinity to nuclear factors from activated macrophages (MURPHY et al. 1995; MA et al. 1996). Interestingly, the phenotype of the ICSBP knockout includes diminished IL-12 production so that perhaps this IRF1 family member may participate in proximal promoter regulation (HOLTSCHKE et al. 1996).

The second potential target of IFN-γ for the above effects would be at the level of the T cell. Thus, under the conditions that were used *in vitro*, it might be necessary for IFN-γ to act on the T cell to enable that cell to respond appropriately to IL-12 for Th1 development (WENNER et al. 1996). This could be mediated by several different mechanisms. In our study comparing the relative contributions of IFN-γ and IFN-α for murine Th1 development (WENNER et al. 1996), we found that IFN-γ alone did not induce Th1 development in vitro, but appeared to promote the responsiveness of T cells to IL-12 during the first 7 days of development. Interestingly, this effect by IFN-γ was not provided by IFN-α.

In the mouse system, we found that IFN-α was unable to induce Th1 development, and was unable to prime T cells to respond to IL-12 as IFN-γ did. It may be important to note that these experiments were done with T cells derived from the BALB/c background and analyzed *in vitro* using the DO11.10 TCR system. T cells from other backgrounds or analyzed in different ways might show differing requirements. However, our conclusion was that IFN-γ clearly acted also at the level of the T cell to enable naive CD4 cells to fully respond to IL-12. Since one potential mechanism of this action would be to induce expression of IL-12 receptors (IL-12Rs), we subsequently addressed these possibilities (see below).

3 The Interleukin-12 Signaling Pathway Is a Point of Regulation for Stabilizing Developing T Helper Phenotype Responses

Studies of the IFN-γ gene have identified several *cis* elements within the promoter and the intragenic regions (SICA et al. 1992; PENIX et al. 1993), but none have yet been identified as playing a direct role in the Th1-specific expression of this gene. Our own attempts to use the IFN-γ gene regulation as a starting point to define Th1-specific signals were not successful. When IFN-γ signaling was shown to operate through the JAK/STAT pathway (SHUAI et al. 1992; FU et al. 1992; SCHINDLER et al. 1992; DARNELL et al. 1994), we set out to ask if IL-12, which we knew induced Th1 development in the mouse, could act through distinct members of the STAT pathway. Therefore, we tried to identify IL-12 induced EMSA complexes using probes known to bind Stat1 (JACOBSON et al. 1995). We found that Stat3 and Stat4, in addition to Stat1, were activated by murine IL-12 in a Th1 clone. While several cytokines can activate either Stat1 or Stat3, Stat4 was apparently activated by no other cytokine than IL-12. Furthermore, at the time we did these studies, there had been no identification of any factor that activated Stat4, which had been identified solely through its homology to Stat1 (ZHONG et al. 1994; YAMAMOTO et al. 1994) and was therefore an orphan STAT. We showed that IL-12 treatment of Th1 cells led to the rapid phosphorylation of Stat4, and that there was a predominant complex apparently composed of a Stat4 homodimer, with the other complexes in the IL-12-treated extracts being Stat1 homodimer and Stat3/Stat4 heterodimers.

Turning back to the IFN-γ promoter, we found that the proximal IFN-γ promoter failed to compete in gel shift assays for the IL-12 induced complexes (JACOBSON et al. 1995). This indicated to us that this region of the IFN-γ promoter did not contain high-affinity sites capable of interacting with IL-12-induced complexes. Subsequently, a study by XU et al. (1996) showed that Stat4 has the ability to form higher order complexes, through NH_2- terminal interactions between Stat4 dimers, to allow Stat4 interactions with lower affinity STAT sites. While this study showed that the intragenic regions of IFN-γ contain several potential low-affinity STAT sites, it did not test whether these sites actually participate functionally in IL-12-induced or Th1-selective expression of IFN-γ. Because we present evidence that differentiated Th1 cells can express high levels of IFN-γ transcripts independently of Stat4 activation (see below), we suggest that other factors (which may themselves be induced by Stat4) may be responsible for the high level of IFN-γ production by Th1 cells.

Knowing that IL-12 induces Stat4 in a specific manner, we next asked if this pathway was regulated during Th1/Th2 development. We compared the ability of Th2 cells and Th1 cells to activate complexes in response to several cytokines (SZABO et al. 1995). While IFN-α and IL-4 can activate STAT-containing complexes similarly in both Th1 and Th2 cells, we found that signaling for IL-12 and for IFN-γ differed between Th1 and Th2 cells. First, only Th1 cells appeared to induce complexes in response to IL-12. Th2 cells failed to induce complexes in response to IL-12, and this appeared to be a developmental process that occurred very rapidly following activation of primary cells. Even within 3 days following activation of FACS-sorted naive T cells, there was a substantial decrease in the ability of IL-12 to induce Stat4 phosphorylation. This capacity was completely lost by day 5 and continued to be absent thereafter.

At the time of this study, the molecular basis for this extinction of IL-12 signaling was unknown, so that it was only possible to demonstrate that high-affinity binding sites were apparently equivalent for both Th1 and Th2 cells. Subsequently we have learned that there are two subunits comprising functional Il-12Rs in both mouse and human (CHUA et al. 1994, 1995; PRESKY et al. 1996; WU et al. 1996). The $\beta1$ subunit, cloned first by Gubler and colleagues (CHUA et al. 1995) in the mouse system, can confer substantial high-affinity binding of the IL-12 heterodimer. This fact explains our observation that both Th1 and Th2 cells appeared to have high-affinity binding sites.

Gubler and colleagues subsequently cloned a second component of the IL-12 receptor, called the ß2 subunit, that is responsible for STAT recruitment and signaling in response to IL-12 (PRESKY et al. 1996). We examined Th1 and Th2 cells for expression of the signaling subunit of the IL-12R. First, Th2 cells clearly lack expression of this signaling subunit. Second of all, this subunit is expressed in Th1 cells, and it appears to undergo specific regulation by the cytokines IL-4, IFN-γ, and IL-12 (SZABO et al. 1997). In the mouse system, it appears that TCR activation is necessary to induce expression of the subunits of the IL-12R, which are absent on naive T cells (SZABO et al. 1997). Induction of the ß2 subunit is inhibited by IL-4, leading to its extinction during Th2 development. However, and somewhat surprisingly, we found that IFN-γ can induce expression of the signaling subunit even when T cells are cultured with IL-4 and undergo development of an IL-4 producing phenotype. These effects were mediated at the level of transcription of the IL-12R ß2 subunit. This observation explains our previous finding that IFN-γ appeared necessary for promoting IL-12 responsiveness. Since we had used T cells from the BALB/c background in our earlier studies, the IL-4 produced by these T cells was apparently sufficient to impose a repression of the ß2 subunit, a process which can be overcome with IFN-γ.

4 Roles of Stat1 and Stat4 in Th1/Th2 Development: Separation of Distinct Phases of Development

We have examined these issues in greater detail by crossing the DO11.10 TCR transgenic model onto Stat1- and Stat4-deficient backgrounds and analyzing Th1/Th2 development *in vitro* under a variety of specific conditions. First, we confirm the finding that Stat4 is required for Th1 development. Stat4-deficient T cells make extremely little IFN-γ under any circumstances. Normal Th1 populations induced in the absence of IL-4 and in the presence of IL-12 can produce, on day 7, IFN-γ in the range of >1000U/ml under our standard in vitro assay system. However, identically treated Stat4-deficient T cells only make between 10 and 20U/ml. This level is above the threshold of detection and can be regulated by IL-12 and by IFN-γ exposure, so that we feel this very small level does represent a weak acquired developmental stage (JACOBSON et al., in preparation). It is perhaps a semantic issue whether these cells should be considered Th1 cells, since while they do not make IL-4, their IFN-γ production is clearly far less than normal Th1 cells. Nonetheless, they produce no IL-4 (if IL-4 has been neutralized in the primary) and do produce some IFN-γ (although the amount is unlikely to be effective against Leishmania). Further experiments are required to fully resolve this issue.

While Stat4 is required to allow IL-12 to augment IFN-γ production, it appears that Stat1 is required to provide for expression of the IL-12R β2 subunit. This conclusion stems from analysis of the pattern of responses of Stat1-deficient

DO11.10 T cells to cytokines *in vitro*. In Stat1-deficient T cells, we found that Th1 development in response to IL-12 could occur as efficiently as in normal T cells, provided that the IL-12 receptors were present. In the absence of Stat1, however, IFN-γ is not able to induce receptor expression as in wild-type cells. IL-12R β2 expression in Stat1-deficient T cells requires IL-4 neutralization and IL-12 addition very early in development. Neutralizing IL-4 alone does not induce Th1 development. All of these findings are consistent with the previous findings in which IL-4 inhibited IL-12R β2 and IFN-γ induced IL-12R β2 expression. Thus we suggest that Th1 development can be considered to be at least a two stage process (Fig. 1). First, after initial TCR activation, the balance of IL-4 and IFN-γ can act to determine whether T cells will express sufficient IL-12 β2 subunit to allow for IL-12 responsiveness. Second, IL-12 , through a mechanism involving Stat4, acts to promote a process of development that provides for a significantly increased capacity for IFN-γ production.

5 Effects of IFN-α on Th1/Th2 Development: Differences Between Human and Murine Signaling

Dr. Francesco Sinigaglia at the same time was carrying out parallel work on IL-12R regulation in the human system (ROGGE et al. 1997), and thus we had the

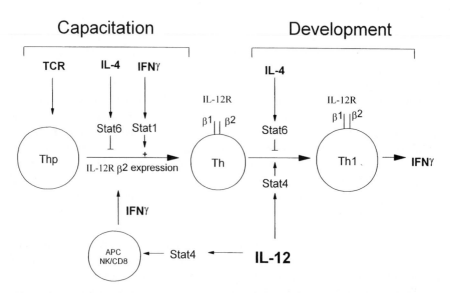

Fig. 1. Th1 development can be considered to proceed through several stages, based on the dependence on specific STAT factors, here focusing on Stat1 and Stat4. The repression of interleukin (IL)-12 receptor expression by IL-4 is implied in the diagram to involve Stat6, but this hypothesis needs to be tested directly

opportunity to compare and contrast our findings (SZABO et al. 1997) with the human system. In the mouse system Th2 cells appear to completely extinguish IL-12 responsiveness through the complete lose of ß2 expression (SZABO et al. 1997). This may not be the case for human Th2 clones, since these cells appear to maintain some low level of ß2 expression, and therefore can respond to IL-12 treatment, activating Stat1, 3 and 4, which then can further induce the expression of this receptor. In contrast, in the mouse cells, sufficiently low levels of receptor are expressed by Th2 cells so that IL-12 treatment alone cannot achieve induction of the ß2 receptor. A second difference between mouse and human was noted: IFN-γ treatment of developing murine Th2 cells appeared capable of inducing IL-12R expression and functional IL-12 responsiveness. However, in the human system, IFN-γ does not act as an effective inducer of IL-12R β2 as it was in the mouse. IFN-α, in contrast, was effective at inducing expression of the β2 subunit in human T cells. Thus, whatever the basis of this difference turns out to be, it may imply that there are significant differences in the pattern of cytokine responses between human and mouse for control of the IL-12 receptor signaling chains.

Furthermore, there has recently been one report that might indicate a difference between IFN-α signaling between mouse and human for the pattern of STATs activated (CHO et al. 1996). This finding, by O'Shea and colleagues, describes Stat4 activation to occur in human peripheral T cell blasts in response to treatment with IFN-α. It was already known that IL-12 activated Stat4 in both mouse and human T cells (JACOBSON et al. 1995; BACON et al. 1995). Because Stat4 appears to have a central role in inducing Th1 development in the mouse, based upon the knockout phenotype, this might predict that IFN-α might be able to induce Th1 development in human T cells. In fact, there has been some uncertainty in the human system whether IL-12 is the only factor able to induce Th1 development, with some reports indicating an effect of IFN-α (BRINKMANN et al. 1993; PARRONCHI et al. 1992).

We re-examined the actions of IFN-α in the mouse system and have reconfirmed the fact that, in the murine system, IFN-α does not activate Stat4, nor does it induce Th1 development (WENNER et al. 1996). Our positive controls for the ability of IFN-α to act on mouse T cells was demonstrated by changes in class I MHC expression and the ability of IFN-α to activate Stat1 (WENNER et al. 1996). We repeated the experiments of O'Shea and confirm their findings, that IFN-α activates Stat4 in human T cells (FARRAR and MURPHY, in preparation). Thus, there appears to be a real difference between human and mouse for the pattern of STAT activation. The basis for this interesting and potentially important difference is unknown at present. This difference could either be at the level of the receptor structure or due to potential differences in STAT structure, since for IFN-α, some STAT activation may rely on recruitment by Stat2 to the activated receptor complex. This will clearly be an active area for current and ongoing work, since the relative actions of IFN-α and IL-12 have important implications for immunotherapy of infectious diseases in humans.

6 Regulation of Interferon-γ Signaling in Developing Th1 and Th2 Populations

We had earlier described a difference in IFN-γ signaling between Th1 and Th2 cells (Szabo et al. 1995). IFN-γ treatment of Th2 cells resulted in the formation of complexes in gel shift from extracts of Th2 cells, but not from extracts of Th1 cells. This finding, in our hands, appeared likely to be due to the high levels of exposure of Th1 cells to IFN-γ, rather than a difference between Th1 and Th2 cells based upon differentiation. We based this conclusion on an experiment in which Th2 cells were activated in the presence of high levels of IFN-γ (Szabo et al. 1995). We noted that Th2 cells thus exposed to IFN-γ also became unresponsive to IFN-γ (Szabo et al. 1995). Therefore we suspected that this phenomenon might be some form of ligand induced down-regulation of IFN-γ signaling, not a key signature of Th1 cells.

To study the molecular basis of this difference, we collaborated with Dr. Robert Schreiber, who developed a number of important reagents to analyze IFN-γ signaling. The results of these studies indicated that the loss of IFN-γ signaling seen in Th1 cells is due to a down-regulation of the ß chain of the IFN-γ receptor, which is essential for Stat1 recruitment and phosphorylation (Bach et al. 1995). Thus, down-regulation of this subunit leads to the inability to signal for IFN-γ responses in either Th1 or Th2 cells. The significance of this down-regulation has not been tested in vivo. This phenomenon was first described by Fitch and Gajewski, who found that there was a significant anti-proliferative response of IFN-γ exerted on Th2 cells, but not on Th1 cells (Gajewski and Fitch 1988; Gajewski et al. 1988, 1989), and who suggested that this may indicate that IFN-γ would play an important role in skewing the phenotype of a population of emerging T cells based upon differential proliferative responses of distinct subpopulations. Since that time, other factors have been found that direct phenotype development in quantitatively greater amounts. Moreover, since we have shown that the mechanism of this difference relies upon exposure of T cells to IFN-γ, it is important to realize that, in vivo, T cells of different antigen specificity reacting to different pathogens may be exposed to the common levels of extracellular IFN-γ and therefore would fall under the same effects equally for down-regulation of the IFN-γ signaling pathway. Therefore, this pathway has yet to be verified as functionally operative *in vivo*.

7 Genetic Effects on Early Th1/Th2 Development

Another parameter that can effect Th1 and Th2 development is genetic background. For example, certain pathogen susceptibilities may involve components of Th1/Th2 phenotype responses, but the genetic basis for the whole *in vivo* phenotype may include more than the T cell axis, and may be multigenic and complex. We have tried to begin analyzing the complex genetic aspects of T helper development by using the DO11.10 TCR transgene bred onto distinct H-2^d genetic backgrounds, allowing us to compare T cells of identical antigen specificity but on distinct genetic backgrounds (HsIEH et al. 1995). We found that B10, BALB/c, and DBA2 genetic backgrounds respond similarly to strong inducing stimuli of IL-12-or IL-4 for Th1 or Th2 development, but that there was a difference between these backgrounds *in vitro* when T cells were allowed to develop without direct manipulation of cytokines, termed operationally as the neutral condition, or perhaps less precisely a default pathway. These terms are not meant to imply that there is a specific or unique default mechanism programmed to operate differently from the known actions of other factors, but simply to provide a term for the conditions that seems to bring out the difference between genetic backgrounds. We understand that *in vitro* conditions are artificial, but we have traded this for the opportunity to examine the genetic basis of differing T cell responses between various strains.

T cells from the B10.D2 genetic background appear to more strongly develop towards a Th1 phenotype when developing under neutral conditions than T cells from a BALB/c background (HsIEH et al. 1995). This difference is maintained independently of the genetic background of the APCs used to prime the T cells. Thus, we seem to be identifying differences residing within the T cell compartment that tend to promote Th1 development in the B10 greater than in the BALB/c background. The basis for this difference appears to be complex. First, while IL-4 induced Th2 development in the B10 and the BALB/c T cells, we did observe residual IFN-γ produced by B10 T cells but not BALB/c T cells. In contrast, neutralization of IL-4 and addition of IL-12 led to nearly identical Th1 phenotypes developing from both backgrounds. The largest difference was in the neutral point, at which B10 T cells were significantly higher in IFN-γ production and BALB/c T cells significantly higher in IL-4 production.

To begin to understand the cellular mechanisms at work, we carried out a series of mixing experiments (GULER et al. 1996). The purpose of these was to distinguish whether genetic effects acted by induction of distinct extracellular conditions in primary cultures, or through cell intrinsic differences, (i.e., such as variations in an intracellular signaling pathway). For example, if BALB/c T cells simply were programmed to produce greater quantities of IL-4 on primary stimulation compared to B10, we would expect a mixing experiment to reveal this difference, since IL-4 produced by BALB/c T cells can act on both the BALB/c and B10.D2 T cells in a mixing experiment. The results from these mixing experiments indicated that both intrinsic and extrinsic differences exist between the B10 and BALB/c cells (GULER et al. 1996).

Another component we analyzed in these mixing experiments was the maintenance of IL-12 responsiveness, since we had recently found that this component of T cell response determined by expression of the ß2 subunit of the IL-12 receptor, was expressed independently of strict Th1/Th2 phenotype definitions. Therefore, we carried out a series of mixing experiments in which phenotype and IL-12 responsiveness were measured in two ways: (1) induction of IFN-γ on secondary stimulation, and (2) induction of an IL-12 responsive cell surface marker, CD25.

The results of these experiments indicated that both an intrinsic and extrinsic difference exist between BALB and B10 T cells. The intrinsic difference appears to allow B10 T cells to maintain IL-12 responsiveness, even when mixed in a population of BALB/c T cells where BALB/c cells predominate. In this setting, B10 remained more Th1-like, producing higher IFN-γ, but more significantly remained IL-12 responsive, whereas BALB/c lose IL-12 responsiveness under these conditions. In the reverse setting, where B10 are mixed at a majority, both B10 and BALB retain an IL-12 responsive phenotype, and BALB/c T cells show more of a Th1 type bulk phenotype pathway, producing higher IFN-γ levels than before.

We have measured expression of the mRNA for the two IL-12R subunits in this system. We find that the maintenance of IL-12 responsiveness at neutral conditions shown by B10 T cells correlates with expression of the IL-12R ß2 subunit, while loss of IL-12 responsiveness in BALB/c correlates with its loss (GULER et al. 1997). Moreover, the maintenance of IL-12R ß2 expression in the B10 appears regulated by cytokines generally as in the BALB/c, since it is induced by IFN-γ and inhibited by IL-4. At present we do not understand the underlying reasons for this difference. When we examined the pattern of inheritance of IL-12 responsiveness at neutral conditions, we found only one locus, near the IL-4 gene on mouse chromosome 11, that was strongly linked to this phenotype (GORHAM et al. 1996). That might imply that the IL-4 locus itself is different between these strains in a manner that preferentially skews BALB/c towards Th2 development more strongly than B10.D2, or that some other gene resides in this region that is responsible for the observed phenotype. The results of the mixing experiments (GULER et al. 1996) seem to favor the second possibility, since the first mechanism would have predicted that mixing BALB/c and B10.D2 cells together would convert B10 cells toward IL-12 unresponsiveness, given that IL-4 acts outside the cells. However, that study only examined the F1(B10.D2 × BALB/c) back-cross to BALB/c, and not a full F2 cohort of mice for genetic analysis, and this may have led us to miss loci that could act in a recessive manner.

8 Summary

Much of our focus in understanding Th1/Th2 development has been on the signals delivered by IL-12 and IL-4 as final determinants of terminal T cell differentiation. Because extinction of IL-12 signaling in early Th2 development could potentially be important in imprinting a more permanent Th2 phenotype on a population of T cells, we have also examined various parameters regulating the IL-12 signaling pathway. Whereas IL-4 appears to repress functional IL-12 signaling through inhibition of IL-12R ß2 expression, IFN-γ in the mouse, and IFN-α in the human appear to induce IL-12R ß2 expression and promote IL-12 responsiveness. We propose that Th1 development can be considered in two stages, capacitance and development. Capacitance would simply involve expression of IL-12R ß1 and ß2 subunits, regulated by TCR, IL-

4 and IFNs. The second stage, development, we propose is the true IL-12 induced developmental stage, involving expression of Stat4 inducible proteins. In the human, this may also occur via IFN-α, which is able to activate Stat4. It is perhaps possible that all of Stat4 actions on Th1 development may be exert directly by Stat4 at the IFN-γ gene, however we suggest that, more likely, Stat4 may act to induce Th1 development through the induction of other non-cytokine genes, whose stable expression maintains the transcriptional state of a Th1 cell.

References

Bach EA, Szabo SJ, Dighe AS, Ashkenazi A, Aguet M, Murphy KM, Schreiber RD (1995) Ligand-induced autoregulation of IFN-gamma receptor beta chain expression in T helper cell subsets. Science 270:1215–1218

Bacon CM, Petricoin EF III, Ortaldo JR, Rees RC, Lamer AC, Johnston JA, O'Shea JJ (1995) Interleukin 12 induces tyrosine phosphorylation and activation of STAT4 in human lymphocytes. Proc Natl Acad Sci USA 92:7307–7311

Belosevic M, Finbloom DS, Van der Meide PH, Slayter MV, Nacy CA (1989) Administration of monoclonal anti-IFN-gamma antibodies in vivo abrogates natural resistance of C3H/HeN mice to infection with Leishmania major. J Immunol 143:266–274

Betz M, Fox BS (1990) Regulation and development of cytochrome c-specific IL-4-producing T cells. J Immunol 145:1046–1052

Brinkmann V, Geiger T, Alkan S, Heusser (1993) Interferon alpha increases the frequency of interferon gamma-producing human CD4+ T cells. J Exp Med 178:1655–1663

Chang T-L, Shea CM, Urioste S, Thompson RC, Boom WH, Abbas AK (1990) Heterogeneity of helper/inducer T lymphocytes. III. Responses of IL-2 and IL-4-producing (Th1 and Th2) clones to antigens presented by different accessory cells. J Immunol 145:2803–2808

Chatelain R, Varkila K, Coffman RL (1992) IL-4 induces a Th2 response in Leishmania major-infected mice. J Immunol 148:1182–1187

Cho SS, Bacon CM, Sudarshan C, Rees RC, Finbloom D, Pine R, O'Shea JJ (1996) Activation of stat4 by IL-12 and IFN-alpha – evidence for the involvement of ligand-induced tyrosine and serine phosphorylation. J Immunol 157:4781–4789

Chua AO, Chizzonite R, Desai BB, Truitt TP, Nunes P, Minetti LJ, Warrier RR, Presky DH, Levine JF, Gately MK et al (1994) Expression cloning of a human IL-12 receptor component. A new member of the cytokine receptor superfamily with strong homology to gp130. J Immunol 153:128–136

Chua AO, Wilkinson VL, Presky DH, Gubler U (1995) Cloning and characterization of a mouse IL-receptor-beta component. J Immunol 155:4286–4294

Cleveland MG, Gorham JD, Murphy TL, Tuomanen E, Murphy KM (1996) Lipoteichoic acid preparations of gram-positive bacteria induce interleukin-12 through a CD14-dependent pathway. Infect Immun 64:1906–1912

Constant S, Pfeiffer C, Woodard A, Pasqualini T, Bottomly K (1995) Extent of T cell receptor ligation can determine the functional differentiation of naive CD4+ T cells. J Exp Med 182:1591–1596

Darnell JE Jr, Kerr IM, Stark GR (1994) Jak-STAT pathways and transcriptional activation in response to IFNs and other extracellular signaling proteins (review). Science 264:1415–1421

Freeman GJ, Boussiotis VA, Anumanthan A, Bemstein GM, Ke XY, Rennert PD, Gray GS, Gribben JG, Nadler LM (1995) B7-1 and B7-2 do not deliver identical costimulatory signals, since B7-2 but not B7-1 preferentially costimulates the initial production of IL-4 . Immunity 2:523–532

Fu XY, Schindler C, Improta T, Aebersold R, Darnell JE Jr (1992) The proteins of ISGF-3, the interferon alpha-induced transcriptional activator, define a gene family involved in signal transduction. Proc Natl Acad Sci USA 89:7840–7843

Gajewski TF, Fitch FW (1988) Anti-proliferative effect of IFN-gamma in immune regulation. I. IFN-gamma inhibits the proliferation of Th2 but not Th1 murine helper T lymphocyte clones. J Immunol 140:4245–4252

Gajewski TF, Goldwasser E, Fitch FW (1988) Anti-proliferative effect of IFN-gamma in immune regulation. 11. IFN-gamma inhibits the proliferation of murine bone marrow cells stimulated with IL-3, IL-4 , or granulocyte-macrophage colony stimulating factor. J Immunol 141:2635 2642

Gajewski TF, Joyce J, Fitch FW (1989) Antiproliferative effect of IFN-gamma in immune regulation. III. Differential selection of TH1 and TH2 murine helper T lymphocyte clones using recombinant IL-2 and recombinant IFN-gamma. J Immunol 143:15 22

Gorham JD, Guler ML, Steen RG, Mackey AJ, Daly MJ, Frederick K, Dietrich WF, Murphy KM (1996) Genetic mapping of a murine locus controlling development of T helper 1 T helper 2 type responses. Proc Natl Acad Sci USA 93:12467 12472

Guler ML, Gorham JD, Hsieh CS, Mackey AJ, Steen RG, Dietrich WF, Murphy KM (1996) Genetic susceptibility to Leishmania: IL-12 responsiveness in TH1 cell development. Science 271:984 987

Guler ML, Jacobson NG, Gubler U, Murphy KM (1997) T cell genetic background influences IL-12 signaling: effects on BALB/c and B10.D2 Th1 phenotype development (abstract) (submitted)

Heinzel FP, Sadick MD, Holaday BJ, Coffman RL, Locksley RM (1989) Reciprocal expression of interferon gamma or interleukin 4 during the resolution or progression of murine leishmaniasis. J Exp Med 169:59 72

Holtschke T, Lohler J, Kanno Y, Fehr T, Giese N, Rosenbauer F, Lou J, Knobeloch KP, Gabriele L, Waring JF, Bachmann MF, Zinkemagel RM, Morse HC, Ozato K, Horak I (1996) Immunodeficiency and chronic myelogenous leukemia-like syndrome in mice with a targeted mutation of the ICSBP gene. Cell 87:307 317

Hosken NA, Shibuya K, Heath AW, Murphy KM, O'Garra A (1995) The effect of antigen dose on CD4+ T helper cell phenotype development in a T cell receptor-alpha beta-transgenic model. J Exp Med 182:1579 1584

Hou J, Schindler U, Henzel WJ, Ho TC, Brassuer M, McKnight SL (1994) An interleukin-induced transcription factor: IL-4 Stat. Science 265:1701 1706

Howard JG (1986) Immunological regulation and control of experimental Leishmaniasis. Int Rev Exp Pathol 28:79 116

Hsieh CS, Macatonia SE, O'Garra A, Murphy KM (1995) T cell genetic background determines default T helper phenotype development in vitro. J Exp Med 181:713 721

Jacobson NG, Szabo SJ, Weber-Nordt RM, Zhong Z, Schreiber RD, Darnell JE Jr, Murphy KM (1995) Interleukin 12 signaling in T helper type 1 (Th1) cells involves tyrosine phosphorylation of signal transducer and activator of transcription (STAT)3 and Stat4. J Exp Med 181:1755 1762

Kaplan MH, Schindler U, Smiley ST, Grusby MJ (1996a) Stat6 is required for mediating responses to IL-4 and for development of Th2 cells. Immunity 4:313 319

Kaplan MH, Sun YL, Hoey T, Grusby MJ (1996b) Impaired EL-responses and enhanced development of Th2 cells in Stat4-deficient mice. Nature 382:174 177

Kopf M, Le Gros G, Bachmann M, Lamers MC, Bluethmann H, Kohler G (1993) Disruption of the murine IL-4 gene blocks Th2 cytokine responses. Nature 362:245 247

Kubin M, Kamoun M, Trinchieri G (1994) Interleukin 12 synergizes with B7/CD28 interaction in inducing efficient proliferation and cytokine production of human T cells. J Exp Med 180:211 222

Kuchroo VK, Das MP, Brown JA, Ranger AM, Zamvil SS, Sobel RA, Weiner HL, Nabavi N, Glimcher LH (1995) B7-1 and B7-2 costimulatory molecules activate differentially the Th1/Th2 developmental pathways: application to autoimmune disease therapy. Cell 80:707 718

Kuhn R, Rajewsky K, Muller W (1991) Generation and analysis of interleukin deficient mice. Science 254:707 710

Le Gros G, Ben-Sasson SZ, Seder RA, Finkelman FD, Paul WE (1990) Generation of interleukin 4 (IL-4)-producing cells in vivo and in vitro: IL-2 and IL-4 are required for in vitro generation of IL-4-producing cells. J Exp Med 172:921 929

Lenschow DJ, Ho SC, Sattar H, Rhee L, Gray G, Nabavi N, Herold KC, Bluestone JA (1995) Differential effects of anti-B7-1 and anti-B7-2 monoclonal antibody treatment on the development of diabetes in the nonobese diabetic mouse. J Exp Med 181:1145 1155

Locksley RM, Scott P (1991) Helper T-cell subsets in mouse leishmaniasis: induction, expansion and effector function. In: Ash C, Gallagher RB (eds) Immunoparasitology Today. Elsevier Trends Journals, UK, a58 a61

Ma X, Chow JM, Gri G, Carra G, Gerosa F, Wolf SF, Dzialo R, Trinchieri G (1996) The interleukin 12 p40 gene promoter is primed by interferon gamma in monocytic cells. J Exp Med 183:147 157

Macatonia SE, Hsieh CS, Murphy KM, O'Garra A (1993) Dendritic cells and macrophages are required for Th1 development of CD4+ T cells from alpha beta TCR transgenic mice: IL-12 substitution for macrophages to stimulate IFN-gamma production is IFN-gamma-dependent. Int Immunol 5:1119–1128

Maggi E, Parronchi P, Manetti R, Simonelli C, Piccini M-P, Rugiu RS, De Carli M, Ricci M, Romagnani S (1992) Reciprocal regulatory effects of IFN-gamma and IL-4 on the in vitro development of human Th1 and Th2 clones. J Immunol 148:2142–2147

Magram J, Connaughton SE, Warrier RR, Carvajal DM, Wu CY, Ferrante J, Stewart C, Sarmiento U, Faherty DA, Gately MK (1996) IL-12-deficient mice are defective in IFN gamma production and type I cytokine responses. Immunity 4:471–471

Manetti R, Parronchi P, Giudizi MG, Piccinni M-P, Maggi E, Trinchieri G, Romangnani S (1993) Natural killer cell stimulatory factor (interleukin 12 [IL-12]) induces T helper type 1 (Th1)-specific immune responses and inhibits the development of IL-4-producing Th cells. J Exp Med 177:1199–1204

Mattner F, Magram J, Ferrante J, Launois P, Di Padova K, Behin R, Gately MK, Louis JA, Alber G (1996) Genetically resistant mice lacking interleukin-12 are susceptible to infection with Leishmania major and mount a polarized Th2 cell response. Eur J Immunol 26:1553–1559

Murphy EE, Terres G, Macatonia SE, Hsieh CS, Mattson J, Lanier L, Wysocka M, Trinchieri G, Murphy K, O'Garra A (1994) B7 and interleukin 12 cooperate for proliferation and interferon gamma production by mouse T helper clones that are unresponsive to B7 costimulation. J Exp Med 180:223–231

Murphy TL, Cleveland MG, Kulesza P, Magram J, Murphy KM (1995) Regulation of interleukin 12 p40 expression through an NF-kappa B half-site. Mol Cell Biol 15:5258–5267

Murray JS, Pfeiffer C, Madri J, Bottomly K (1992) Major histocompatibility complex (MHC) control of CD4 T cell subset activation. II. A single peptide induces either humoral or cell-mediated responses in mice of distinct MHC genotype. Eur J Immunol 22:559–565

Parish CR, Liew FY (1972) Immune response to chemically modified flagellin. 3. Enhanced cell-mediated immunity during high and low zone antibody tolerance to flagellin. J Exp Med 135:298–311

Parronchi P, De Carli M, Manetti R, Simonelli C, Sampognaro S, Piecinni MP, Macchia D, Maggi E, del Prete G, Romagnani S (1992) IL-4 and IFN (alpha and gamma) exert opposite regulatory effects on the development of cytolytic potential by Th1 or Th2 human T cell clones. J Immunol 149:2977–2983

Penix L, Weaver WM, Pang Y, Young HA, Wilson CB (1993) Two essential regulatory elements in the human interferon gamma promoter confer activation specific expression in T cells. J Exp Med 178:1483–1496

Pfeiffer C, Murray J, Madri J, Bottomly K (1991) Selective activation of Th1 and Th2-like cells in vivo-Response to human collagen IV. Immunol Rev 123:65–84

Presky DH, Yang H, Minetti LJ, Chua AO, Nabavi N, Wu CY, Gately MK, Gubler U (1996) A functional interleukin 12 receptor complex is composed of two beta-type cytokine receptor subunits. Proc Natl Acad Sci USA 93:14002–14007

Quelle FW, Shimoda K, Thierfelder W, Fischer C, Kim A, Ruben SM, Cleveland JL, Pierce JH, Keegan AD, Nelms K et al (1995) Cloning of murine Stat6 and human Stat6, Stat proteins that are tyrosine phosphorylated in responses to IL-4 and IL-3 but are not required for mitogenesis. Mol Cell Biol 15:3336–3343

Rogge L, Barberis-Maino L, Biffi M, Passini N, Presky DH, Gubler U, Sinigaglia F (1997) Selective expression of an interleukin-receptor component by human T helper I cells. J Exp Med 185:825 (abstract)

Sadick MD, Heinzel FP, Holaday BJ, Pu RT, Dawkins RS, Locksley RM (1990) Cure of murine leishmaniasis with anti-interleukin 4 monoclonal antibody. J Exp Med 171:115–127

Schindler C, Fu XY, Improta T, Aebersold R, Darnell JE Jr (1992) Proteins of transcription factor ISGF-3: one gene encodes the 91- and 84-kDa ISGF-3 proteins that are activated by interferon alpha. Proc Natl Acad Sci USA 89:7836–7839

Scott P, Natovitz P, Coffman RL, Pearce E, Sher A (1988) Immunoregulation of cutaneous Leishmaniasis. T cell lines that transfer protective immunity or exacerbation belong to different T helper subsets and respond to distinct parasite antigens. J Exp Med 168:1675–1684

Seder RA, Gazzinelli R, Sher A, Paul WE (1993) Interleukin 12 acts directly on CD4 + T cells to enhance priming for interferon gamma production and diminishes interleukin 4 inhibition of such priming. Proc Natl Acad Sci USA 90:10188–10192

Shimoda K, van Deursen J, Sangster MY, Sarawar SR, Carson RT, Tripp RA, Chu C, Quelle FW, Nosaka T, Vignali DA, Doherty PC, Grosveld G, Paul WE, Ihle JN (1996) Lack of IL-4-induced Th2 response and IgE class switching in mice with disrupted Stat6 gene. Nature 380:630–633

Shuai K, Schindler C, Prezioso VR, Darnell JE Jr (1992) Activation of transcription by IFN-gamma: tyrosine phosphorylation of a 91-kD DNA binding protein. Science 258:1808 1812

Sica A, Tan TH, Rice N, Kretzschmar M, Ghosh P, Young HA (1992) The c-rel protooncogene product c-Rel but not NF-kappa B binds to the intronic region of the human interferon-gamma gene at a site related to an interferon-stimulable response element. Proc Natl Acad Sci USA 89:1740 1744

Swain SL, Huston G, Tonkonogy S, Weinberg AD (1991) Transforming growth factor-beta and IL-4 cause helper T cell precursors to develop into distinct effector helper cells that differ in lymphokine secretion pattern and cell surface phenotype. J Immunol 147:2991 3000

Swain SL, Weinberg AD, English M, Huston G (1990) IL-4 directs the development of Th2-like helper effectors. J Immunol 145:3796 3806

Sypek JP, Chung CL, Mayor SEH, Subramanyam D, Goldman SJ, Sieburth DS, Wolf SF, Schaub RG (1993) Resolution of cutaneous Leishmaniasis: interleukin-12 initiates a protective T helper type 1 immune response. J Exp Med 177:1797 1802

Szabo SJ, Dighe AS, Gubler U, Murphy KM (1997) Regulation of the ILR beta2 subunit expression in developing Th1 and Th2 cells. J Exp Med 185:817 (abstract)

Szabo SJ, Jacobson NG, Dighe AS, Gubler U, Murphy KM (1995) Developmental commitment to the Th2 lineage by extinction of IL-12 signaling. Immunity 2:665 675

Takeda K, Tanaka T, Shi W, Matsumoto M, Minami M, Kashiwamura S, Nakanishi K, Yoshida N, Kishimoto T, Akira S (1996) Essential role of Stat6 in IL-4 signalling. Nature 380:627 630

Thierfelder WE, van Deursen JM, Yamamoto K, Tripp RA, Sarawar SR, Carson RT, Sangster MY, Vignali DA, Doherty PC, Grosveld GC, Ihie JN (1996) Requirement for Stat4 in interleukin-mediated responses of natural killer and T cells. Nature 382:171 174

Wenner CA, Guler ML, Macatonia SE, O'Garra A, Murphy KM (1996) Roles of IFN-gamma and IFN-alpha in IL-12-induced T helper cell-1 development. J Immunol 156:1442 1447

Wu CY, Warrier RR, Carvajal DM, Chua AO, Minetti LJ, Chizzonite R, Mongini PKA, Stem AS, Gubler U, Presky DH, Gately MK (1996) Interleukin-12, interleukin-12 receptor, Interleukin-12 receptor beta, and chain. Biological function and distribution of human interleukin-receptor beta chain. Eur J Immunol 26:345 350

Xu X, Sun YL, Hoey T (1996) Cooperative DNA binding and sequence selective recognition conferred by the STAT amino-terminal domain (see comments). Science 273:794 797

Yamamoto K, Quelle FW, Thierfelder WE, Kreider BL, Gilbert DJ, Jenkins NA, Copeland NG, Silvennoinen O, Ihle JN (1994) Stat4, a novel gamma interferon activation site-binding protein expressed in early myeloid differentiation. Mol Cell Biol 14:4342 4349

Zhong Z, Wen Z, Damell JE Jr (1994) Stat3 and Stat4: members of the family of signal transducers and activators of transcription. Proc Natl Acad Sci USA 91:4806 1480

Redirecting Th2 Responses in Allergy

P. Parronchi, E. Maggi, and S. Romagnani

1	Introduction	27
2	Definition and Properties of Th2 Cells	28
3	Mechanisms Involved in the Regulation of Th2 Cell Development	30
3.1	IL-4 Dictates Th2 Cell Differentiation	30
3.1.1	Role for FcεR¹ Non-T Cells	30
3.1.2	Role for NK1.1¹ T Cells	31
3.1.3	Autocrine IL-4 Production by Naive T Cells	31
3.2	Intracellular Signaling for Th2 Cell Development	32
3.3	Cytokines Dampening IL-4 Production and Th2 Cell Development	33
4	Mechanisms Possibly Responsible for Allergen-Specific Th2 Responses in Atopic Subjects	35
4.1	Genetic Factors	35
4.2	Environmental Factors	37
5	Possible Novel Immunotherapeutic Strategies for Atopic Diseases	39
5.1	Induction of Anergy in Allergen-Specific Th2 Cells	40
5.2	Redirection of Allergen-Specific Th2 Responses	41
5.2.1	Th1-Inducing Cytokines	42
5.2.2	Altered Peptide Ligands	42
5.2.3	Recombinant Microorganisms and Other Adjuvants Incorporating Allergen Peptides	43
5.2.4	Plasmid DNA (Allergen Epitope) – Gene Therapy	44
5.2.5	CpG Dinucleotides	45
5.3	Targeting Th2 Cells or Th2-Dependent Effector Molecules	45
5.3.1	Targeting Transcription Factors Related to Th2 Cells	46
5.3.2	Antagonizing IL-4	46
5.3.3	Antagonizing IL-5	46
5.3.4	Targeting IgE	46
6	Concluding Remarks and Summary	47
	References	48

1 Introduction

Atopic diseases are genetically determined disorders which affect approximately 20%–30% of the general population in developed countries and whose prevalence is increasing over the last decades. They are characterized by an enhanced ability of

Istituto di Medicina Interna e Immunoallergologia, University of Florence, Viale Morgagni 85, Florence 50134, Italy

B lymphocytes to produce IgE antibodies in response to certain groups of otherwise innocuous environmental antigens that can activate the immune system after inhalation or ingestion, and perhaps after penetration through the skin (allergens). Allergen-specific IgE antibodies bind indeed to high affinity (type I) Fcε receptors (FcεRI) present on the surface of mast cells/basophils and allergen-induced FcεRI cross-linking results in the release of vasoactive mediators, chemotactic factors and cytokines that in turn trigger the "allergic cascade" (LICHTENSTEIN 1993). In addition to IgE-producing B cells and IgE-binding mast cells/basophils, eosinophils are also involved in the pathogenesis of atopic diseases, inasmuch as these cells accumulate in the sites of allergic inflammation and the toxic products they release significantly contribute to the induction of tissue damage (GLEICH et al. 1994). In the past few years, the cellular basis for the regulation of IgE antibody production has been discovered. More importantly, the intense study on the functional properties of helper T (Th) cells that collaborate with B cells for the synthesis of IgE antibodies has allowed clarification of the mechanisms accounting for the joint involvement of IgE-producing B cells, mast cells/basophils and eosinophils.

This novel view, which has been defined the "Th2 hypothesis in allergy," suggests that atopy is a Th2-driven hypersensitivity to innocuous antigens (allergens) of complex genetic and environmental origins (ROMAGNANI 1994a; 1995a). The considerable progress in our understanding has provided the basis for the development of novel therapeutic strategies of atopic diseases (ROMAGNANI 1995a). The data supporting the validity of the Th2 hypothesis in allergy are summarized in Table 1.

Table 1. Evidence suggesting a pathogenic role for allergen-specific Th2 cells in allergic disorders

---

- Allergens preferentially expand Th cells showing a Th2-like profile (WIERENGA et al. 1990; PARRONCHI et al. 1991)
- Th2-like cells accumulate in the target organs of allergic patients (MAGGI et al. 1991; VAN DER HEIJDEN et al. 1991; HAMID et al. 1991; ROBINSON et al. 1992)
- Allergen challenge results in local activation and recruitment of allergen-specific Th2-like cells (DEL PRETE et al. 1993a; BURASTERO et al. 1995,
- Allergen-reactive Th2 cells expressing membrane CD30, an activation marker preferentially associated with the production of Th2-type cytokines, are present in the circulation of allergic patients during seasonal allergen exposure (DEL PRETE et al. 1995a)
- Successful specific immunotherapy associates with the down-regulation of allergen-reactive Th2 cells and/or up-regulation of allergen-reactive Th1 cells (VARNEY et al. 1993; SECRIST et al. 1993; JUTEL et al. 1995; MCHUGH et al. 1995)

We review the results of our and other studies on the properties of Th2 cells, the mechanisms responsible for their development, and the possibility to redirect the Th2 response in subjects suffering from atopic disorders.

2 Definition and Properties of Th2 Cells

The concept that a particular subset of T cells is responsible for the joint involvement of IgE-producing B cells, mast cells/basophils and eosinophils in the pathogenesis of allergic reactions have become clearer after the demonstration that polarized forms of the specific immune response, based on their profile of cytokine secretion, exist not only in mice (MOSMANN et al. 1986), but also in humans (DEL PRETE et al. 1991; ROMAGNANI 1991). Type 1 helper CD4[+] T cells (Th1) secrete (IL-2), tumor necrosis factor (TNF)-β and IFN-γ and are the principal effectors of phagocyte-mediated host defense, which is highly protective against infections sustained by intracellular parasites (ROMAGNANI 1995b, 1996a). Type 2 helper CD4[+] T cells (Th2), on the other hand, produce IL-4 and IL-13 which stimulate IgE and IgG1 antibody production, IL-5 (an eosinophil-activating factor), and IL-10, which together with IL-4 and IL-13 inhibit several macrophage functions. Therefore, Th2 cells are mainly responsible for phagocyte-independent host defense, e.g., against certain gastrointestinal nematodes (ROMAGNANI 1995b, 1996a), and are excellent candidates to explain why the mast cell/eosinophil/IgE-producing B cell triad is involved in the pathogenesis of allergy (ROMAGNANI 1994a,b, 1995a, 1997a,b).

Although human T-cell clones with characteristics similar to murine Th1 and Th2 clones exist as well (DEL PRETE et al. 1991; ROMAGNANI 1991), so strict a dichotomy including all cytokines as described in mice is rarely found in humans. Therefore a more realistic definition may be that Th2 cells are CD4[+] effector T cells which produce IL-4 but not IFN-γ (ROMAGNANI 1994b, 1996, 1997a,b). More recently, based on the demonstration that Th2 cells can maintain the potential to develop into IFN-γ producers, whereas vigorously primed Th1 cells fail to develop into IL-4 producers, another definition of Th1/Th2 cells has been proposed. Th2 cells are CD4[+] T cells which produce IL-4 while Th1 cells are CD4[+] T cells which cannot (HU-LI et al. 1997). Whatever definition is used, the Th1/Th2 model provides an interesting paradigm for the study and understanding of several pathophysiological processes and possible for the development of novel immunotherapeutic strategies.

Based on the above findings that Th1 and Th2 cells are not originally distinct Th cell subsets but represent polarized forms of the response which develop under the mixed action of genetic and environmental factors (see below), it is highly improbable that they express clearly distinct and selective surface markers. However, some molecules which are preferentially associated with Th1 or Th2 cells have been described. For example, CD30 and CD62L are expressed prevalently during Th2 responses (reviewed in ROMAGNANI 1997). CD30 is associated mainly with Th2 cells since its expression is IL-4 dependent in both mice (NAKAMURA et al. 1997) and humans (ANNUNZIATO et al 1997).

More recently the association of some chemokine receptors with Th2 cells has also been reported. The eotaxin receptor CCR3, which is expressed mainly by eosinophil and basophil granulocytes, has been claimed to be a selective marker of Th2 vs. Th1 cells (SALLUSTO et al. 1997). Likewise, CCR4 (the TARC and MDC receptor) and CCR8 (the I309 and vMIP-II receptor) were found to be expressed, even at a higher extent than CCR3, on human Th2 cells (BONECCHI et al. 1998; ZINGONI et al. 1998). Finally, CXCR4 (receptor for SDF-1 and coreceptor for T-tropic HIV-1 strains), is up-regulated by IL-4 and down-regulated by IFN-γ on both naive and memory T cells (JOURDAN et al. 1998; GALLI et al. 1998). The molecular mechanisms responsible for these preferential associations still need to be clarified. Nevertheless, a flexible program in the expression of some surface molecules by human Th1 or Th2 cells seems to exist and may play a relevant role in numerous biological processes, including allergic inflammation.

3 Mechanisms Involved in the Regulation of Th2 Cell Development

3.1 IL-4 Dictates Th2 Cell Differentiation

The mechanisms responsible for the development of Th2 cells have extensively been investigated. The presence of IL-4 at the site of antigen presentation appears to be the most dominant factor in determining the likelihood for Th2 polarization of the naive Th cell in both mice and humans (SWAIN et al. 1990; SEDER et al. 1992a; MAGGI et al. 1992; SORNASSE et al. 1996). Hormonal factors, such as progesterone, and the CD30L/CD30 interaction can also contribute to the development of human Th2-like cells at least in vitro (PICCINNI et al. 1995; DEL PRETE et al. 1995b), but their in vivo importance in this process remains to be established. Major candidates for IL-4 production at the onset of an immune response include FcεR[+] non-T cells, a specialized T-cell subset corresponding to the murine CD4[+] NK1.1[+] T cells, or the naive Th cell itself.

3.1.1 Role for FcεR[+] Non-T Cells

Several types of FcεR[+] non-T cells are able to produce IL-4. First, murine mast cell lines capable of producing IL-4 have been described (PLAUT et al. 1989). Then, non-T, non-B cells from mouse spleen and human bone marrow, probably belonging to the mast cell/basophil lineage were found to be able to synthesize IL-4 (BEN SASSON et al. 1990; PICCINNI et al. 1991). The IL-4-producing capacity of mouse non-T cells expands dramatically in *Nippostrongylus brasiliensis* (*N. brasiliensis*) infection and in association with anti-IgD injection (CONRAD et al. 1990), suggesting that these cells participate in lymphokine production in helminthic infections and in other situations marked by striking elevations of serum IgE levels. These non-T/non-B cells represent the dominant source of IL-4 and IL-6 in the spleens of immunized animals. Exposing these cells to antigen-specific IgE or IgG in vivo (or in vitro) "armed" them to release both IL-4 and IL-6 upon subsequent antigenic challenge (AOKI et al. 1995). Accordingly, both human mast cells and basophils produce IL-4 in response to several secretagogues (BRADDING et al. 1992; BRUNNER et al. 1993; MACGLASHAN et al. 1994; OKAYAMA et al. 1995). More recently, it has been shown that activated human eosinophils can also release high IL-4 concentrations (MOQBEL et al. 1995; NONAKA et al. 1995; NAKAJIMA et al. 1996). Thus, at least potentially, FcεR[+] non-T cells might very well reflect a means through which Th2 cells can be strikingly amplified in vivo during allergic reactions and parasitic infestations. However, it is unlikely that parasites or allergens would be able to cross-link their receptors prior to a specific immune response that had produced parasite-specific IgG and IgE antibodies. A way out of this dilemma may be a pathway of IL-4 secretion independent from FcεR cross-linking. It has been suggested that both helminth products and some allergens may induce FcεR⁺ cells to release IL-4 because of their proteolytic activity (FINKELMAN et al. 1991a; DUDLER et al. 1995) or via the induction of C5a (OCHENSBERGER et al. 1995). However, obvious mechanisms for FcεR-independent IL-4 production for the great majority of allergens or helminth components have not been identified yet. On the other hand, mast cell-deficient mice develop normal Th2 responses (WERSHILL et al. 1994). Finally, in IL-4-deficient mice only those mice which are reconstituted with IL-4-producing T (but not with IL-4-producing non-T) cells produce antigen-specific IgE (SCHMITZ et al. 1994).

3.1.2 Role for NK1.1⁺ T Cells

CD4⁺NK1.1⁺ T cells represent a specialized subset of T cells which are selected by the nonpolymorphic MHC class I molecule CD1 (BENDELAC et al. 1995). The role of these cells in favoring the development of Th2 cells by providing IL-4 at the onset of an immune responses is suggested by several findings. First, splenic CD4⁺NK1.1⁺ cells are able to secrete large and transient amounts of IL-4 mRNA as soon as 90 min after intravenous injection of anti-CD3 antibody (YOSHIMOTO et al. 1994). Moreover, β_2-microglobulin (β_2M)-deficient mice and SJL mice, which contain a few or no CD4⁺NK1.1⁺ T cells, appear to be severely affected in their capacity for rapid induction of IL-4 mRNA in response to anti-CD3, as well as in their ability to synthesize IgE (YOSHIMOTO and PAUL 1995a, b). Finally, transgenic mice overexpressing CD4⁺NK1.1⁺ T cells exhibit increased IL-4 and IgE production (BENDELAC et al. 1996). However, these MHC class I-dependent CD4⁺NK1.1⁺ cells that give a rapid IL-4 response in the spleen do not appear to contribute significantly to early induced IL-4 responses in the lymph nodes (VAN DER WEID et al. 1996). In addition, studies investigating *Leishmania major* (*L. major*) infection and responses to other pathogens or antigens were unable to incriminate these cells in Th2 responses (LAUNOIS et al. 1995; GUÉRY et al. 1996; BROWN et al. 1996). Finally, it is unlikely that all antigens that promote the differentiation of naive Th cells into the Th2 pathway are capable of activating CD4⁺NK1.1⁺, CD1-restricted, T cells. An alternative possibility may be that another subset of NK1.1⁺ cells, the CD4 CD8 $\gamma\delta^+$ subset, can ensure high levels of IL-4 at the priming (VICARI et al. 1996). The selection of these cells is indeed independent of β_2M-associated class I molecule expression, because they are present even in β_2M-deficient mice (VICARI et al. 1996).

3.1.3 Autocrine IL-4 Production by Naive T Cells

A final possibility is that the maturation of naive T cells into the Th2 pathway mainly depends upon the levels and kinetics of IL-4 production by naive T cells themselves at priming. This possibility was strongly supported by several find-ings obtained in both mice and humans. First, low intensity signaling of T cell receptor (TCR), such as that mediated by low peptide doses or by mutant peptides, led to secretion of low levels of IL-4 by murine naive T cells (PFEIFFER et al. 1995; CONSTANT and BOTTOMLY 1997). Naive T cells, recently activated in the presence of costimulatory molecules-expressing fibroblasts (in the absence of outside influences from other cells), required two or more stimulation events to produce IL-4 and IL-5. This induction of Th2-type cytokine secretion was blocked by inhibiting IL-4 action, which suggests a role for endogenous IL-4 produced by the naive T cells themselves (CROFT and SWAIN 1995). Likewise, human CD45RA [+] (naive) adult peripheral blood T cells, as well as human neonatal T cells, were found to develop into IL-4-producing cells in the absence of any pre-existing source of IL-4 and despite the presence of anti-IL-4 anti-bodies (KALINSKI et al. 1995; DEMEURE et al. 1995; YANG et al. 1995). Finally, high proportions of T cell clones showing a clear-cut Th2 profile of cytokine production could be generated from CD4 [+] αβ [+] T cells isolated from thymus of small children (MINGARI et al. 1996). A significant fraction of uncommitted T cells may be primed for a Th2 phenotype independent of antigen and IL-4 if they are exposed to IL-2 and simultaneously interact with accessory cells bearing the natural CD28 ligands B7-1 and B7-2. When stimulated by specific antigen, such primed Th2 precursor cells may provide a source of IL-4 to promote Th2 immunity (BRINKMANN et al. 1996).

Thus, the maturation of naive T cells into the Th2 pathway can depend upon the levels and the kinetics of autocrine IL-4 production at priming. Obviously, when CD1-restricted antigens are expressed on antigen-presenting cells (APCs), CD4 [+] NK1 [+] T cells which are able to rapidly release high amounts of IL-4 may contribute to the development of the Th2 pathway. Likewise, cytokines from other cells (e.g., mast cell/basophil), hormones, as well as other still unknown microen-vironmental factors, may be involved, suggesting the existence of multiple path-ways in the development of Th2 cells.

3.2 Intracellular Signaling for Th2 Cell Development

The recognition that IL-4 expression during immune responses is critical for determining the development of Th2 cells has intensified interest in the molec-ular basis of its regulation. It is now clear that following the interaction between IL-4 and its receptor on a given cell, the IL-4 induced signal transducer and activator of transcription (STAT) protein, initially designated IL-4 STAT and hereafter termed STAT6, is activated. STAT6 is related in primary amino acid sequence to STAT1, yet encoded by a different gene. The essential role of STAT6 in IL-4 signaling has clearly been demonstrated in STAT6-deficient mice. In these animals T cells are indeed unable to develop into Th2 cells, and pro-duction of IgE and IgG1 is virtually abolished (SHIMODA et al. 1996; TAKEDA et al. 1996). However, there is yet no direct evidence that STAT6 transactivates the IL-4 promoter in T cells, or that the STAT6 site of the IL-4 promoter is required for promoter activity. On the other hand, other transcription factors of the nuclear factor of activated T cells (NF-AT) family are able to transactivate the IL-4 promoter, but they are expressed in both Th1 and Th2 cells (LI-WEBER et al. 1997). Thus it is unlikely that these factors can explain the Th2-specific expression of the IL-4 gene. NIP45 (NF-AT interacting protein) is another factor expressed in Th2 cells which appears to function as a potent coactivator of IL-4 gene transcription (HODGE et al. 1996). However, its expression in Th cells is unclear. By contrast, the proto-oncogene c-maf appears to be expressed selectively in Th2 clones and is induced during Th2, but not during Th1, differentiation (HO et al. 1996). Moreover, its activity appears to be specific to the other Th2 cytokine genes, such as IL-5 and IL-10.

A transcription factor that may be more widely involved in the induction and maintenance of the Th2 pattern of cytokine secretion is GATA-3 (KO et al. 1991). GATA-3 is expressed in both immature and mature T cells and is selectively expressed during Th1, but not during Th2, differentiation (ZHENG and FLAVELL 1997). Thus, in contrast to c-maf, which appears to be IL-4 specific, GATA-3 may function as a more general regulator of Th2 cytokine expression (SZABO et al. 1997).

3.3 Cytokines Dampening IL-4 Production and Th2 Cell Development

Opposite to IL-4, there are other cytokines which are able to down-regulate the development and/or function of Th2 cell. These include IL-12, IFN-γ, and IFN-α.

IL-12, a heterodimeric cytokine produced primarily by phagocytic cells, with multiple activities on NK cells and T cells, including augmentation of IFN-γ production (CHEHIMI and TRINCHIERI 1995), is the dominant factor directing the development of Th1 cells. The Th1-inducing effect of IL-12 was contemporarily and independently demonstrated in both mice and humans (HSIEH et al. 1993; MANETTI et al. 1993, 1994). Recent in vivo data from IL-12-deficient mice show that if Th1 responses were impaired, even if not completely lacking, the magnitude of delayed type hypersensitivity was substantially decreased and secretion of IL-4 was enhanced (MAGRAM et al. 1996; THIERFELDER et al. 1995). IL-12 is able to induce tyrosine phosphorylation and DNA-binding of STAT3 and STAT4 (BACON et al. 1995). STAT-deficient mice have impaired Th1 development and their lymphocytes show enhanced propensity to develop into Th2 cells, as evidenced by increased Th2 cytokine secretion under conditions favoring Th1 or Th2 development (KAPLAN et al. 1996). A locus that controls the maintenance of IL-12 responsiveness, and therefore favors the preferential development of Th1 cells, has recently been described in B10.D2 mice (GORHAM et al. 1996). Of note, this locus (T cell phenotype switch-1 or Tps-1) maps on a region of chromosome 11 which is syntenic with the locus on human chromosome 5q31.1, shown to be associated with elevated serum IgE levels (MARSH et al. 1994) (see above). This suggests that defects in IL-12 expression or IL-12 signaling pathway can lead to a reciprocal up-regulation of Th2 activity. Indeed, besides its well known ability in priming Th cells to produce IFN-γ, IL-12 also inhibits the differentiation of T cells into IL-4-secreting cells, i.e., into Th2 cells (SEDER et al. 1993; MANETTI et al. 1993). This inhibitory effect is, in part, a direct effect (WANG et al. 1994), and in part is mediated by its activity on APCs (DE KRUYFF et al. 1995) and in part by its stimulation of IFN-γ synthesis (SEDER et al. 1993; MANETTI et al. 1993; WYNN et al. 1995). IL-12 is not capable, however, of suppressing a Th2 cell recall response (BLISS et al. 1996). The in vivo effects of IL-12 appear to be, to a large extent, IFN-γ-dependent (MORRIS et al. 1994; WYNN et al. 1995).

The addition of IFN-γ to cultures containing optimal amounts of IL-4 failed to inhibit the priming of CD4[+] T cells from TCR transgenic mice to develop into IL-4-producing T cells. However, when suboptimal concentrations of IL-4 were used for priming, IFN-γ caused a significant decrease in the amount of IL-4 produced after restimulation (SEDER et al. 1992b). This result is consistent with the observation that few IL-4-producing T cell clones emerge from a culture containing IFN-γ (FITCH et al. 1993). IFN-γ can indeed regulate the development of Th1 cells independently of IL-12 (BRADLEY et al. 1996) and, at relatively low concentrations, inhibits the proliferation of murine Th2 clones stimulated with antigen, mitogens, or anti-TCR antibodies; however, secretion of lymphokines, including IL-4, in response to these stimuli is not affected (FITCH et al. 1993). IFN-γ also inhibits proliferation of murine Th2 clones exposed to IL-2 or IL-4 (GAJEWSKI and FITCH 1989). Although not absolute, the inhibitory effect of IFN-γ on proliferation of murine Th2 cells is significant and seems to be sufficient to limit the clonal expansion of such cells. While IFN-γ does not inhibit lymphokine secretion by stimulated Th2 cells, IFN-γ blocks many agonist effects of those secreted cytokines. For example, IFN-γ inhibits proliferation of murine bone marrow cells stimulated with IL-3, IL-4, or granulocyte/macrophage colony-stimulating factor (GM-CSF) (GAJEWSKI and FITCH 1990). In addition IFN-γ can inhibit IL-4-dependent B cell differentiation (ABED et al. 1994). The proliferation of human Th2 clones is also inhibited by IFN-γ (DEL PRETE et al. 1993b).

IFN-α also plays a negative regulatory role on the development of Th2 cells. In particular, in the mouse, IFN-α was found to suppress increases in the level of splenic IL-4 mRNA induced by either treatment with anti-IgD antibody or infection with *N. brasiliensis*, whereas in both these conditions the levels of IFN-γ mRNA were increased (FINKELMAN et al. 1991b; URBAN et al. 1993). In the absence of IFN-γ, IFN-α augments IL-12 effects on the inhibition of subsequent IL-4 production rather than enhancing IL-12 priming for subsequent IFN-γ production (WENNER et al. 1996). Recently, it has been shown that, like IL-12, IFN-α can induce tyrosine phosphorylation and DNA-binding of STAT4 (CHO et al. 1996). The regulatory activity of IFN-α has also been demonstrated in the human system. IFN-α inhibits the development of allergen-specific T cells into Th2-like cells (PARRONCHI et al. 1992b; ROMAGNANI 1992; PARRONCHI et al. 1996), and up-regulates the expression of the IL-12 receptor β chain in naive human T cells (ROGGE et al. 1997).

4 Mechanisms Possibly Responsible for Allergen-Specific Th2 Responses in Atopic Subjects

Based on above findings, it is reasonable to hypothesize that the preferential development of allergen-specific Th2-like responses seen in atopic subjects is determined by: (a) the nature and the intensity of TCR signaling by the allergen peptide ligand, and (b) the altered regulation of IL-4 production by Th cells and/or other cell types, which is probably controlled by both genetic and environmental factors (Fig. 1).

4.1 Genetic Factors

The possibility that atopic subjects have a genetic dysregulation at level of Th cell-derived IL-4 is supported by several observations. First, CD4[+] T cell clones from atopic individuals are able to produce noticeable amounts of IL-4 and IL-5 in response to bacterial antigens, such as PPD and streptokinase, that usually evoke responses with a restricted Th1-like cytokine profile in nonatopic individuals (PARRONCHI et al. 1992a). Second, atopic donors have a higher frequency of IL-4-producing T cells than normal subjects (CHAN et al. 1996). Moreover, T cell clones generated from cord blood lymphocytes of newborns with atopic parents produce higher IL-4 concentrations than neonatal lymphocytes of newborns with nonatopic parents (PICCINNI et al. 1996). Recently, large panels of *Parietaria officinalis* group I (Par O 1)-specific T cell clones were generated from donors with low or high serum IgE levels and assessed for both profile of cytokine production and reactivity to two immunodominant Par O 1 peptides (p92 and p96). Of note, both p92- and p96-specific T cell clones generated from "high IgE" donors produced remarkable amounts of IL-4 and low amounts of IFN-γ. By contrast, the majority of p96-specific T cell clones generated from "low IgE" donors produced high amounts of both IL-4 and IFN-γ, whereas most p92-specific T cell clones generated from the same donors showed a Th1-like profile (high IFN-γ and low IL-4) (P. Parronchi et al., 1998).

Taken together, these data strongly suggest that the allergen peptide ligand can influence the cytokine profile of Th cells, but mechanisms underlying noncognate regulation of IgE responsiveness are overwhelming. These data are also consistent with the results of several studies in which the direct candidate approach has been used to evaluate potential roles for various genes in a number of chromosomal regions in both atopy and asthma.

The HLA region on chromosome 6p21.3 contains an obvious group of candidate immune response (Ir) genes involved in antigen presentation. Over the past several years numerous studies have provided evidence that HLA-D genes are involved in specific immune responsiveness to allergens, and therefore in the expression of specific atopic sensitivities (BARNES and MARSCH 1998). Likewise, significant linkage of IgE responses to certain allergens to the TCRα/β region encoded

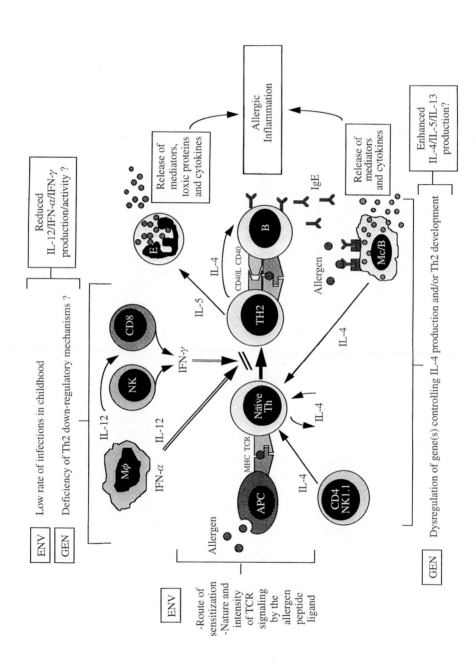

on chromosome 14q11.2-12 has been observed (COOKSON 1998). However, another set of studies relates to evidence that a gene(s) on human chromosome 5q31-q33 exerts a more general control of IgE production and may play a major role in the expression of atopy. Chromosome 5q31-q33 contains multiple candidate genes for atopy, including IL-3, IL-4, IL-5, IL-9, IL-13 and CSF2. The products of these genes are involved in the B-cell isotype switching to IgE and in the up-regulation of eosinophil, basophil, mast cell, and IgE functions, i.e., in the Th2 response. Evidence for linkage with total IgE (not specific IgE) was found within the 5q31.1 region, but not for three markers lying just outside this region in the Pennsylvania Old Order Amish (Marsh et al. 1994). Moreover, linkage of total IgE concentrations for marker D5S436, which maps within about 100 kb of GRL, and about 11 Mb telomeric of IL-4 has been found. These data clearly indicate that genes in the 5q31-33 region are involved in elevated IgE production, with a major candidate gene being IL-4. Other candidate gene(s) for allergy and asthma are present in a large section of chromosome 12q14.3–q24. They include INFG (which inhibits IL-4 activity), SCF (a mast cell growth factor), NFYB (which up-regulates transcription of IL-4 and HLA-D genes), and STAT6 (essential cytokine-regulated transcription; see above). When the potential role of these candidate genes was studied in multiplex Afro-Caribbean "asthma" families from Barbados, evidence was found of linkage with both asthma and IgE concentration (BARNES et al. 1996). Finally, despite several conflicting results, the involvement of a gene(s) in the 11q1–13 region (where the FCεRI is encoded) (COOKSON 1998) and a mutation in the α-subunit of the IL-4 receptor (IL-4R) (HERSHEY et al. 1997) with atopy and IgE production have also been suggested.

4.2 Environmental Factors

The observation that the prevalence of allergy has increased over the last few decades in developed countries (BRUCE et al. 1993; SCHULTZ-LARSEN 1993) clearly suggests that, in addition to genetic alterations possibly at the level of mechanisms governing IL-4 gene expression, environmental factors also contribute to regulate

---

Fig. 1. The pathogenic role of allergen-specific Th2 cells in allergic inflammation and the possible mechanisms responsible for their development in atopic subjects. Ubiquitous allergens stimulate the preferential development of Th2 cells because of either environmental (*ENV*) or genetic (*GEN*) reasons. ENV factors favoring Th2 responses are the route of allergen entry and the nature and intensity of TCR signaling by the allergen peptide ligand, which operate over the entire population. Other ENV factors may be the low rate of infections that induce the production of Th1-type cytokines in early childhood when allergen sensitization occurs. Among GEN factors, dysregulation of gene(s) controlling IL-4 production (which is required for Th2-cell development) and/or a deficiency of mechanisms that favor the production of Th2-inhibitory cytokines may be involved. Allergen-specific Th2 cells produce IL-4 (which promotes IgE switching in B cells), and IL-5 (which promotes differentiation, activation, and in situ survival of eosinophils). Cross-linking by allergen of IgE bound to Fcε receptors on mast cells/basophils induces the release of mediators and cytokines, including IL-4, which in turn may amplify the Th2 response. Mediators, toxic proteins, and cytokines released by both eosinophils and mast cell/basophils initiate and sustain the allergic inflammation

the development of Th2 cells and/or of their function. The environmental factors may act either before or after birth. Some years ago we first showed that the immune response to *Dermatophagoides pteronyssinus* begins during fetal life (PICCINNI et al. 1993), a finding which has then been confirmed by others (JONES et al. 1996; VAN DUREN-SCHMIDT et al. 1997). During pregnancy a Th2-skewed priming probably occurs in all cases, due to the maternal environment. Successful pregnancy may indeed be characterized by a switch from Th1 to Th2 in order to reduce the reactivity of the maternal immune system against the fetal allograft (WEGMANN et al. 1993). Accordingly, progesterone, at the concentrations present at the fetomaternal interface, favors the development of T cells into IL-4 producing cells (PICCINNI et al. 1995). More recently we have found that T-cell clones generated from the decidua of women suffering from unexplained recurrent abortions showed significantly reduced production of IL-4, IL-10, and LIF than T-cell clones generated from the decidua of women with underlying voluntary abortion (normal gestation) (PICCINNI et al. 1998). Thus the pregnancy-related environment may favor a weak Th2-skewed priming to transplacental allergens, which is obviously enhance under the influence of an "atopic" genetic background (PICCINNI et al. 1996).

The environmental factors acting after birth, however, are certainly more important in affecting the individual outcome in the Th response to ubiquitous allergens, and they probably account for the increased prevalence of allergy over recent decades in Western countries. Four years ago, when the increase prevalence of allergy was generally attributed to pollution (MIYAMOTO and TAKAFUJI 1991; KRAMER et al. 1991; RING and BEHRENDT 1993), we proposed the hypothesis that this increase is "at least partly related to the strong reduction in childhood (just when aeroallergen sensitization is occurring) infections, particularly tubercular infection, which previously induced the production of cytokines antagonistic to Th2 development" (ROMAGNANI 1994a). This hypothesis (the "hygiene hypothesis") was based mainly on our in vitro data showing that cytokines produced by "natural immunity" cells (macrophages, NK cells) in response to *Mycobacterium tuberculosis* and/or its components were able to shift, at least in vitro, the development of allergen-specific T cells from the Th2/Th0 to the Th1 profile (ROMAGNANI 1992; MAGGI et al. 1992; PARRONCHI et al. 1992b; MANETTI et al. 1993, 1995). This possibility has recently been substantiated by studies in mice showing that IFN-γ produced during the Th1 immune response against BCG suppresses the development of local inflammatory Th2 responses in the lung (ERB et al. 1998). More importantly, three recent epidemiological studies strongly support the concept that the decline in tubercular infection and perhaps in other infections in childhood is an important factor underlying the rising severity and prevalence of atopic disorders in developed countries over the last decades. A highly significant inverse association between delayed hypersensitivity to *M. tuberculosis* and atopy has indeed been found among Japanese schoolchildren. Positive tuberculin responses predicted a lower incidence of asthma, lower serum IgE levels, and cytokine profiles biased toward Th1 responses, suggesting that exposure and response to *M. tuberculosis* during childhood may, by modification of immune profiles, inhibit the development of atopic disorders (SHIRAKAWA et al. 1997). Likewise, African children who had been vaccinated with measles virus showed significantly increased risk of prick test positivity to house dust mite than nonvaccinated children who had had natural measles infection (SHAHEEN et al. 1996). Finally, hepatitis A seropositivity is inversely associated with atopy among Italian military students (MATRICARDI et al. 1997). When the possible role of various types of infection was compared, the food-borne *Toxoplasma gondii* and *Helicobacter pylori*, hepatitis A appeared to be protective towards the development of atopy, whereas viruses transmitted mainly by other routes (measles, mumps, rubella, chicken-pox, cytomegalovirus, and herpes simplex virus type 1) did not (P.M. Matricardi, personal communication), suggesting that the pattern of orally administered microbes rather than any type of infection regulates the risk of developing atopy. Major causes of changes in infectious diseases and in the pattern of microbial exposure which are associated with Westernization include cleaner water and food, less crowded accommodation, changes in natural history of exposure due to better nutrition, vaccinations, and antimicrobial treatments.

5 Possible Novel Immunotherapeutic Strategies for Atopic Diseases

The insights into the pathogenesis of allergic disorders are providing exciting opportunities for the development of novel immunomodulatory regimens. The new approaches are directed to target allergen-specific T cells (allergen-specific immunotherapy) or their effector molecules (nonallergen-specific immunotherapy). Allergen-specific immunotherapy may consist of induction of unresponsiveness in T cells and/or prevention or redirection of Th2 responses. Non-allergen-specific immunotherapy may be based on antagonization of Th2 cytokines and on anti-IgE therapies (Table 2).

Table 2.

Allergen-Specific Immunotherapy
– Induction of anergy in Th2 cells by allergen-derived peptides
– Redirection of allergen-specific Th2 responses
– Allergen plus Th1-inducing cytokines
– Altered allergen peptide ligands
– Allergen peptides incorporated in recombinant microorganisms or appropriate adjuvants
– Plasmid DNA (allergen epitope) gene therapy
Non-Allergen-Specific Immunotherapy
– Targeting selective Th2 transcription factors (c-Maf, STAT6)
– Targeting IL-4 (soluble IL-4 receptors; IL-4 mutant protein)
– Targeting IL-5 (humanized anti-IL-5 antibody; inhibitors of IL-5 transcription,
– Targeting IgE ("intelligent" humanized anti-IgE antibody)

5.1 Induction of Anergy in Allergen-Specific Th2 Cells

Anergy, or the induction of a state of functional nonresponsiveness, in allergen-specific Th2 cells was achieved at least in vitro by incubation of *Dermatophagoides pteronyssinus* group 1 (Der p1)-specific Th2 clones with high doses of the relevant Der p1-derived peptides in the absence of APCs (FASLER et al. 1995). A similar approach had previously been used on influenza virus-specific human T cell clones incubated with the relevant virus-specific hemagglutinin peptide (LAMB et al. 1987). Under these experimental conditions, both influenza- and Der p1-specific T cell clones lost the ability to respond to subsequent stimulation with the intact antigen and APCs (FASLER et al. 1995; LORD and LAMB 1996), as well as to provide B cell help for IgE production even in the presence of exogenous IL-4 and IL-13 (FASLER et al. 1995). At present, however, there is limited information on the possibility to tolerize allergen-specific Th2 responses in vivo. The activation of allergen-specific T cells and immunoglobulin synthesis in mice can be inhibited in vivo by intranasal, oral or subcutaneous administration of Der p1 or certain Der p1 peptides (HOYNE et al. 1995). Subcutaneous immunization with a dominant peptide of the major cat allergen *Felix domesticus* group 1 (Fel d1) resulted in decreased T cell responses, as measured by reduced IL-2 production to subsequent challenge with this peptide in Fel d1-primed mice. Furthermore, pretreatment of mice with two Fel d1-derived immunodominant peptides induced T cell tolerance to a subsequent challenge with the entire recombinant Fel d1 chain, suggesting that peptides containing some of the T cell epitopes in a large polypeptide can tolerize T cells for subsequent challenge with that larger polypeptide protein (BRINER et al. 1993). Finally, rush immunotherapy with an equimolar mixture of three phospholipase A_2 (PLA) peptides appeared to be protective against live bee sting and resulted in the induction of T cell tolerance towards the entire PLA antigen, which could be reversed by in vitro treatment of cells with IL-2 and IL-15 (ADKIS et al. 1996).

Thus, the above studies support the possibility to use peptide-induced T cell anergy to treat human allergic diseases. The major advantage of this type of specific immunotherapy in comparison with the classic therapy with crude extracts or purified native allergens is its significant potential safety. Binding of allergen-specific IgE to allergen-derived peptides is strongly reduced (VAN'T HOFF et al. 1991), with the desired effects on T cells unaltered. However, some conceptual and practical problems still need to be overcome. First, although peptides containing minimal T cell epitopes cannot cause anaphylactic reactions, the possibility that high amounts of IL-4 and IL-5 are produced during initiation of peptide vaccination, thus causing initial worsening of the allergic reactions, cannot be excluded. More importantly, most allergens contain multiple T cell epitopes which are restricted by different MHC class II antigens. Thus, although some studies have shown that total T cell responses to complex native proteins are indeed limited to a few immunodominant epitopes (GAMMON and SERCARZ 1989; BRINER et al. 1993), the possibility exists that T cells can escape the induction of tolerance in the presence of other, nonimmunodominant T cell epitopes, which become available after in vivo processing of native allergen (GAMMON and SERCARZ 1989). This problem, however, may be circumvented by administering mixtures of overlapping allergen-derived peptides representing all T cell activation-inducing epitopes on the corresponding native allergen, thus avoiding both mapping of all T cell activation-inducing epitopes and precise MHC class II typing of patients.

A clinical trial, based on vaccinating cat allergic patients with immunodominant peptides derived from the major cat allergen Fel d1 has recently been performed. The preliminary results show that the treatment is safe and may result in clinical improvement (NORMAN et al. 1994). Likewise, a trial with bee venom PLA-derived peptides has shown good clinical efficacy with no adverse reactions (MULLER et al. 1997).

5.2 Redirection of Allergen-Specific Th2 Responses

Another strategy for the neutralization of allergen-specific Th2 responses would be to change the cytokine profile of allergen-specific CD4$^+$ T cells. One possibility to induce a T cell class switch in the cytokine profile may be to act directly on allergen-specific Th2 cells. This strategy raises the question of the stability of established Th2 effectors. Once established, the Th2 phenotype appears to be more stable than the Th1 phenotype in both mice and humans. Fully polarized murine Th2 cells are resistant to a reversal in their cytokine production patterns, which may reflect the lack of receptors for Th1 induction on these cells (PEREZ et al. 1995; SZABO et al. 1995). Likewise, the activation of established Th2-like human T cell clones in the presence of IL-12 can result in the transient expression of very low amounts of IFN-γ, which is lost if the stimulation is repeated in the absence of IL-12 (MANETTI et al. 1994). Strongly polarized allergen-specific human Th2 clones are even unable to respond to IL-12, as shown by the total lack of IL-12-induced phosphorylation of STAT4 (HILKENS et al. 1996). Thus, it is reasonably clear that reversibility of polarized Th2 effectors is lost after long-term stimulation (MURPHY et al. 1996).

The possibility to influence the development of memory Th cells seems more likely. Initial studies suggested that once naive T cells were committed to Th1 or Th2 patterns of cytokine secretion even their memory populations would secrete identical arrays of cytokine upon antigen restimulation (SWAIN et al. 1990; LE GROS et al. 1993). However, more recent findings indicate that regardless of prior commitment, memory CD4$^+$ T cells may retain the capacity to be further influenced during effector cell development to become subsets that are at least temporarily polarized to a particular pattern of cytokine secretion (MOCCI and COFFMAN 1995; BRADLEY et al. 1996). For example, administration of IL-12 to mice reverses detrimental *L. major*-specific Th2 responses, when given with a drug which reduces the parasite load (NABORS et al. 1995). These results support our previous findings on human Th subsets, suggesting that memory cells previously committed in vivo to the Th2 pattern of cytokine secretion can be induced to produce cytokines of the opposite phenotype by restimulation in vitro in the presence of IFN-γ and anti-IL-4 antibody, IFN-α, or IL-12 (MAGGI et al. 1992; PARRONCHI et al. 1992b; MANETTI et al. 1993; PARRONCHI et al. 1996). Accordingly, human neonatal T cells primed to

Th2 cell phenotype by IL-4 were not stable and could rapidly be reverted into a population predominantly containing Th0 and Th1 cells after a single restimulation in the presence of IL-12 (SORNASSE et al. 1996). In addition, it has been shown that successful specific immunotherapy in vivo is associated with changes in the cytokine profile of allergen-reactive Th2 cells (VARNEY et al. 1993; SECRIST et al. 1993; JUTEL et al. 1995; McHUGH et al. 1995). Thus, a potentially successful approach would be to prime memory allergen-specific Th cells in a manner which selects for the prevalent Th1 phenotype. Indeed, predominant secretion of Th1-type cytokines in response to allergen will not only result in development and recruitment of different effector responses (IgG production and macrophage activation instead of IgE production and eosinophil activation), but will also probably inhibit the function of established Th2 responses as a result of cross-regulatory circuits (see above).

5.2.1 Th1-Inducing Cytokines

Up-regulation of allergen-specific Th1 responses was achieved in vitro by using cytokines, such as IFN-γ, IFN-α (MAGGI et al. 1992; PARRONCHI et al. 1992b; PARRONCHI et al. 1996) and IL-12 (MANETTI et al. 1993). It was subsequently demonstrated that the Th1-inducing effect of IL-12 was mainly due to its ability to prime T cells to IFN-γ production (MANETTI et al. 1994). A similar effect was observed by adding to bulk peripheral blood mononuclear cell (PBMC) culture an antibody directed against the signaling lymphocyte activation molecule (SLAM) (DE VRIES 1996), a novel 70 kDa T cell surface receptor capable of inducing or enhancing IFN-γ production by human T cells (COCKS et al. 1995). Up-regulation of allergen-specific Th1 cell development was also obtained with polyI:C, which induced high level production of both IFN-α and IL-12 by macrophages (MANETTI et al. 1995b). Interestingly, IFN-α not only modulated the cytokine profile of allergen-specific T cells, but it also affected their TCR repertoire (PARRONCHI et al. 1996).

The regulatory effects of Th1-inducing cytokines on allergen-specific responses have been demonstrated also in vivo. Mice lacking the IFN-γ receptor have impaired ability to resolve a lung eosinophilic inflammatory response associated with a prolonged capacity of T cells to exhibit a Th2 cytokine profile (COYLE et al. 1996). Moreover, nebulized IFN-γ decreases IgE production and normalizes airway function in a murine model of allergen sensitization (LACK et al. 1994). IFN-α inhibits the eosinophil and CD4[1] T cell infiltration when induced into the trachea of ovalbumin (OVA)-immunized mice (NAKAJIMA et al. 1994). IL-12 can block antigen-induced airway hyperresponsiveness and pulmonary inflammation in vivo by suppressing Th2 cytokine expression (GAVETT et al. 1995). These data suggest that the injection of selected allergen peptides in combination with Th1-inducing cytokines may represent the basis for a novel immunotherapeutic strategy in allergic disorders.

5.2.2 Altered Peptide Ligands

Several experimental models suggest that certain epitopes can preferentially induce one of the two subsets of Th cells. A repetitive peptide of *L. major* selectively activated Th2 cells and enhanced disease progression (LIEW et al. 1990). By contrast, a recombinant 403 amino acid protein from *L. brasiliensis*, homologous to ribosomal protein eIF4A (LeIF), preferentially stimulated human PBMCs to express a Th1-type cytokine profile and to produce IL-12 (SKEIKY et al. 1995). Changes as little as those concerning single amino acid residues greatly deviated the ability of the immunodominant CD4 T cell epitope of the bacteriophage λ cI repressor protein to induce IL-4 production and immediate-type hypersensitivity (SOLOWAY et al. 1991). Likewise, a peptide differing from its wild type in a single residue failed to evoke a proliferative response by a Th2 clone, but retained the ability to induce IL-4 production (EVAVOLD et al. 1992; RACIOPPI et al. 1993). Varying either the antigenic peptide or the MHC class II molecules could determine whether Th1-like or Th2-like responses are obtained. High MHC class II peptide density on the APC surface favored Th1-like responses, while low ligand densities favored Th2-like responses (PFEIFFER et al. 1995). Likewise, by using a set of ligands with various class binding affinities but unchanged T cell specificity, it was shown that stimulation with the highest affinity ligand resulted in IFN-γ production. In contrast, ligands that demonstrated relatively lower MHC class II binding induced only IL-4 secretion (KUMAR et al. 1995). A single MHC polymorphism may dictate Th1/Th2 selection by determining the level of peptide presented to a given TCR on APCs (MURRAY et al. 1995).

Recently, the p28–40 analogues with alanine residues at positions 34 and 36 of the dominant T cell epitope of the group 2 mite allergen were found able to alter the ratio of IFN-γ/IL-4 production in human Th0 cells by selectively enhancing IFN-γ secretion (TSITOURA et al. 1996). Likewise, replacement of the 21st residue arginine by lysine in the peptide fragment [18]RSLRTVTPIRMQGG[31], derived from a *D. farinae* group 1 allergen, resulted in a significant increase in IFN-γ production, with remarkable changes in IL-4 production, by a specific human Th0 clone, which was also associated with increased production of IL-12 (MATSUOKA et al. 1996).

5.2.3 Recombinant Microorganisms and Other Adjuvants Incorporating Allergen Peptides

In general, it appears that corpuscular immunogens more easily promote Th1 responses than soluble antigens. The Th1-inducing activity of corpuscular immunogens is probably related to their greater ability to induce IL-12 production by macrophages that are responsible for phagocytosis. Recombinant *M. bovis* (BCG) strains have been proposed as optimal adjuvants for the induction of Th1 responses against *L. major* (ABDELAK et al. 1995), *M. tuberculosis* (MURRAY et al. 1996) and measles virus (FENNELLY et al. 1995). Orally administered live recombinant *Salmonella* also elicites dominant antigen-specific Th1-type responses in both mucosal and systemic tissues, in the absence of expression of Th2 cytokines IL-4 and IL-5 (VAN COTT et al. 1996). Thus, allergens may be engineered into recombinant *Mycobacteria* or other appropriate microorganisms.

Antigen polymerization also preferentially evokes Th1-type responses in comparison to the unmodified antigen. Following administration of high molecular weight OVA polymers, there was marked inhibition of OVA-specific IgE and strong increase of IgG2a production (HAYGLASS et al. 1991; GIENI et al. 1993). This was due to marked up-regulation of IFN-γ and down-regulation of IL-10 production, whereas the frequency of IL-4-producing cells was unchanged in response to the polymerized form of antigen (GIENI et al. 1996). The type of adjuvant is also important in determining the profile of cytokine secretion. Complete Freund's adjuvant evokes Th1 responses (JANEWAY et al. 1988), probably because the components of the mycobacterial cell wall induce high IL-12 production by macrophages, with subsequent up-regulation of IFN-γ production (see below), whereas incomplete Freund's adjuvant, alum, acellular *Bordetella pertussis* toxin, and cholera toxin expand prevalent Th2 responses (XU-ARMANO et al. 1993; MARINARO et al. 1995; FORSTHUBER et al. 1996). Moreover, influenza virus envelope proteins incorporated in immunostimulating complexes stimulate a prevalent Th1 response, whereas the same antigens in a micelle induce a more prominent Th2 response (VILLACRES-ERIKSSON et al. 1995). Of interest is the recent finding indicating that conjugation of recombinant antigens to the polysaccharide mannan (polymannose) under reducing conditions favors predominant Th2 responses. By contrast, the same antigen coupled to oxidized mannan selectively promotes Th1 responses (APOSTOLOPOULOS et al. 1995).

5.2.4 Plasmid DNA (Allergen Epitope) – Gene Therapy

Another revolutionary approach to the treatment of allergic disorders, originally developed for gene therapy, may be based on vaccination with plasmid vectors. These genetic vaccines are commonly called naked DNA or polynucleotide vaccines. Naked DNA consists of a desired gene inserted into a plasmid which can enter, and remain episomally in, cells close to the injection site, where it is subsequently transcribed and translated, causing expression of the gene product (TANG et al. 1992). There are three major advantages with this technique. First, naked DNA vaccines could bypass the numerous problems associated with other vectors, such as immune responses against the delivery vectors and concern about safety related to the use of any viral vector. Second, they could induce the expression of antigens that resemble native epitopes more closely than do standard vaccines, where the process of manufacturing can alter the structure of the protein and thus lower the antigenicity of the vaccine. Third, they may be constructed so that genes from several different allergens are included on the same plasmid, thus potentially decreasing the number of vaccinations required. Finally and most importantly, the gene protein enters the cells' MHC class I pathway since only proteins that originate inside a cell are processed by this pathway. This results in the stimulation of CD8[+] cytotoxic T cells and evokes cell-mediated immunity. Accordingly, the results of a recent study suggest that intradermal immunization of BALB/c mice with naked plasmid DNA encoding for *Escherichia coli* β-galactosidase (β-gal) induces a Th1 immune response, as demonstrated by a highly restricted IgG2a antibody response to β-gal together with IFN-γ, but not IL-4 or IL-5, secretion by in vitro β-gal-stimulated splenic CD4[+] T cells. Immunization with the corresponding protein, β-gal, in saline or alum induces a Th2 response (RAZ et al. 1995). More-over, intramuscular injection in mice of a gene construct containing an important house dust mite allergen gene (Der p5) resulted in the induction of Der p5-specific IgG antibodies, but not IgE antibody (Hsu et al. 1996). The suppression of specific IgE response could be adaptively transferred with CD8 [+] T cells producing high levels of IFN-γ (Hsu et al. 1996), suggesting that allergen gene transfer is indeed effective in modulation of allergen-specific IgE responses.

5.2.5 CpG Dinucleotides

More recently it has been shown that bacterial DNA contains immunostimulatory motifs consisting of unmethylated CpG dinucleotides, which rapidly trigger an innate immune response characterized by T-cell independent polyclonal B-cell ac-tivation (KRIEG et al. 1995), direct NK cell stimulation (BALLAS et al. 1996), and production of IL-6, IL-12, IL-18, and IFN-α by macrophages and/or dendritic cells (KLINMAN et al. 1997; CHU et al. 1997). Synthetic oligodeoxynucleotides that contain unmethylated CpG motifs also bias the specific immune response to a Th1-dominated cytokine pattern (CHU et al. 1997). How mouse macrophages, B lym-phocytes, and NK cells recognize specific DNA sequences in bacterial DNA is still unclear. The immunostimulatory CpG motifs could theoretically bind to comple-mentary sequences in DNA or mRNA or interact with one or more signal trans-duction molecules in the cytoplasm, or on the plasma membrane. However, if the data generated in murine models prove reproducible in humans without side effects of cytokine overproduction, it is conceivable that the coadministration of an immunostimulatory CpG with a given allergen could be used in atopic persons to modify the Th2-oriented allergen-specific response.

5.3 Targeting Th2 Cells or Th2-Dependent Effector Molecules

Other potentially efficient therapeutic strategies may be based on targeting Th2 cells or the effector molecules produced as a consequence of activation of Th2 cells. This approach has become conceptually acceptable after evidence was accumulated suggesting that Th2 responses are probably not critical for survival and protection. First, IL-4-deficient mice which do not develop Th2 responses are better protected than wild animals against the great majority of infections (KOPF et al. 1993). Th2 cells are indeed more protective than Th1 cells only against gastrointestinal nem-atodes (URBAN et al. 1995); Th2 cells do not kill nematodes, but disturb their habitat via the effects exerted by IL-4 on gastrointestinal function (URBAN et al. 1995). At present, however, gastrointestinal nematodes do not represent a health problem at least in developed countries. Also, it should be noted that subjects homozygous for deletions involving the ε gene seem perfectly normal (BRUSCO et al. 1995), suggesting that at least one of the most important IL-4-dependent effector mechanisms (production of IgE antibody) is not critical. Thus, targeting of either

Th2 cells or Th2-dependent effector molecules may be a reasonable form of immunotherapy in patients with severe atopic disorders.

5.3.1 Targeting Transcription Factors Related to Th2 Cells

The demonstration that some transcription factors are critical for IL-4 production and/or Th2 cell development (see above), such as the c-*maf* oncogene product and STAT6, suggests that they may be appropriate targets for manipulating Th2 responses. The c-*maf* oncogene is indeed selectively transcribed in Th2 cells and appears to control tissue expression of IL-4 (Ho et al. 1996). On the other hand, signaling by IL-4 occurs through activation of STAT6, and knock-out of the *stat* 6 gene results in deficient Th2 responses (SHIMODA et al. 1996; TAKEDA et al. 1996).

5.3.2 Antagonizing IL-4

The activity of IL-4 may be antagonized by soluble IL-4 receptors (RENZ et al. 1995) and even more effectively by a human IL-4 mutant protein (AVERSA et al. 1993). Interestingly, this protein also antagonizes the biological activity of IL-13 as well as the IL-4-driven differentiation of Th2 cells in vitro (SORNASSE et al. 1996).

5.3.3 Antagonizing IL-5

The activity of IL-5 has been antagonized by humanized antibodies to IL-5, which appeared to be capable of inhibiting eosinophil infiltration and normalizing airway hyperreactivity in monkeys challenged with *Ascaris suis* (EGAN et al. 1995; KUNG et al. 1995). IL-5 activity has also been antagonized by an IL-5-specific gene transcription inhibitor (MORI et al. 1996).

5.3.4 Targeting IgE

"Intelligent" humanized anti-human IgE antibodies have recently been developed. These antibodies bind to IgE residues critical for receptor binding, thus reacting with both membrane IgE from B cells and soluble IgE, but not with mast cell- or basophil-bound IgE (JARDIEU 1995; COYLE et al. 1996). Interestingly, one of these types of antibodies not only neutralized serum IgE, but also inhibited both the recruitment of eosinophils into the lungs of house dust mite-sensitized mice and the production of IL-4 and IL-5, but not IFN-γ, by inhibiting IgE-CD23-facilitated antigen presentation to T cells (COYLE et al. 1996). Thus, IgE targeting not only blocks the IgE-mediated allergic inflammation, but also down-regulates Th2 responses and the subsequent infiltration of eosinophils into the airways. Human preclinical safety trials of anti-IgE therapy appear to be satisfactory and clinical studies look very promising (FAHY et al. 1997; CORNE et al. 1997).

6 Concluding Remarks and Summary

In the last few years, evidence has been accumulated to suggest that allergen-reactive Th2 cells play a triggering role in the activation and/or recruitment of IgE antibody-producing B cells, mast cells and eosinophils, the cellular triad involved in allergic inflammation. Th2 cells do represent a polarized arm of the effector specific response that plays some role in the protection against gastrointestinal nematodes. They also act as regulatory cells for chronic and/or excessive Th1-mediated responses. Th2 cells are generated from precursor naive Th cells when the specific antigen is encountered in an IL-4-containing microenvironment. However, the source of IL-4 required at the initiation of the response for the development of naive Th cells into Th2 effectors is still unknown. Both FcεR $^+$ non-T cells and a specialized subset of CD1-restricted NK1.1 $^+$ CD4 $^+$ T cells have been suggested to quickly provide IL-4 to the naive Th cell; however, a more likely possibility is that the maturation into the Th2 pathway mainly depends upon the levels and the kinetics of autocrine IL-4 production by naive Th cells themselves at priming.

The question of how these Th2 responses are enhanced in atopic patients is also unclear. Both the nature of the TCR signaling provided by the allergen peptide ligand and a dysregulation of IL-4 production likely concur to determine the Th2 profile of allergen-specific Th cells, the genetic dysregulation of IL-4 production probably being overwhelming. Some gene products selectively expressed in Th2 cells, such as c-Maf, or selectively controlling the expression of IL-4, such as STAT6, have recently been described. Moreover, CD8 $^+$ T cell subsets and cytokines, such as IL-12, IFN-α, and IFN-γ, that dampen the production of IL-4 as well as the development and/or the function of Th2 cells have been identified. These findings suggest that an inherited dysregulation of gene(s) controlling IL-4 expression and/or alterations of mechanisms that regulate Th2 development and function may favor Th2 responses against common environmental allergens, particularly in patients who develop severe atopic disorders. However, the role of environmental factors influencing the balance between Th1 and Th2 responses in the development of mild or common atopy should also be considered. For example, the significant reduction in childhood infections (particularly tubercular infection) that induce the production of cytokines antagonistic to Th2 development may be partly responsible for the rising prevalence of allergy observed in the last few decades in developed countries.

The new insights into the pathophysiology of T cell responses in atopic diseases provide exciting opportunities for the development of novel immunotherapeutic strategies. These include the induction of nonresponsiveness in allergen-specific Th2 cells by allergen peptides or redirection of allergen-specific Th2 responses by Th1-inducing cytokines, altered peptide ligands, allergens incorporated into recombinant microorganisms or bound to appropriate adjuvants, and plasmid DNA vaccination. In patients with severe atopic disorders, the possibility of nonallergen-specific immunotherapeutic regimens designed to target Th2 cells or Th2-dependent effector molecules, such as specific IL-4 transcription factors, IL-4, IL-5, and IgE, may also be considered.

Acknowledgements. The experiments reported here have been performed with grants provided by CNR, EC, AIRC, and Istituto Superiore di Sanità (AIDS project). The authors wish to thank Roberto Manetti, Francesco Annunziato, and Salvatore Sampognaro for their excellent collaboration.

References

Abdelhak S, Louzir H, Timm J, Blel L, Benlasfar Z, Lagranderie M, Gheorghiu M, Dallagi K, Gicquel B (1995) Recombinant BCG expressing the leishmania surface antigen Gp62 induces protective immunity against *Leishmania major* infection in BALB/c mice. Microbiology 141:1582 1592

Abed NS, Chace JH, Cowdery JS (1994) T cell-independent and T cell-dependent B cell activation increases IFN-gamma receptor expression and renders B cells sensitive to IFN-gamma-mediated inhibition. J Immunol 153:3369 3377

Akdis CA, Akdis M, Blesken T, Wymann D, Alkan S, Muller U, Blaser K (1996) Epitope specific T cell tolerance to phospholipase A2 in bee venom immunotherapy and recovery by IL-2 and IL-15 in vitro. J Clin Invest 98:1676 1683

Aoki I, Kinzer C, Shirai A, Paul WE, Klinman DM (1995) IgE receptor-positive non-B/non-T cells dominate the production of interleukin 4 and interleukin 6 in immunized mice. Proc Natl Acad Sci USA 92:2534 2538

Apostolopoulos V, Pietersz GA, Loveland BE, Sandrin MS, McKenzie IFC (1995) Oxidative/reductive conjugation of mannan to antigen selects for T1 or T2 immune responses. Proc Natl Acad Sci USA 92:10128 10132

Aversa G, Punnonen J, Cocks BG, de Waal Malefyt R, Vega F Jr, Zurawski SM, Zurawski G, de Vries JE (1993) An interleukin 4 (IL-4) mutant protein inhibits both IL-4 or IL-13-induced human immunoglobulin G4 (IgG4) and IgE synthesis and B cell proliferation: support for a common component shared by IL-4 and IL-13 receptors. J Exp Med 178:2213 2218

Bacon CM, Petricoin EF, Ortaldo JE, Rees RC, Larner AC, Johnston JA, O'Shea JJ (1995) Interleukin 12 induces tyrosine phosphorylation and activation of STAT4 in human lymphocytes. Proc Natl Acad Sci USA 92:7307 7311

Ballas ZK, Rasmussen WL, Krieg AM (1996) Induction of NK activity in murine and human cells by CpG motifs in oligogeoxynucleotides and bacterial DNA. J Immunol 157:1840 1845

Ben Sasson SZ, Le Gros G, Conrad DH, Finkelman FD, Paul WE (1990) Cross linking Fc receptors stimulates splenic non-B, non-T cells to secrete interleukin 4 and other lymphokines. Proc Natl Acad Sci USA USA 87:1421 1425

Bendelac A, Lantz O, Quimby ME, Yewdell JW, Bennink JR, Butkiewicz RR (1995) CD1 recognition by mouse NK1.1⁺ T lymphocytes. Science 268:863 865

Bendelac A, Hunziker R, Lantz O (1996) Increased Interleukin 4 and Immunoglobulin E production in transgenic mice overexpressing NK1.1 T cells. J Exp Med 184:1285 1293

Bliss J, Van Cleave V, Murray K, Wiencis A, Katchum M, Maylor R, Haire T, Resmini C, Abbas AK, Wolf SF (1996) IL-12, as an adjuvant, promotes a T helper 1 cell, but does not suppress a T helper 2 cell recall response. J Immunol 156:887 894

Bonecchi R, Bianchi G, Bordignon Panina P, D'Ambrosio D, Lang R, Borsatti A, Sozzani S, Allavena P, Gray PA, Mantovani A, Sinigaglia F (1998) Differential expression of chemokine receptors and chemotactic responsiveness of Th1 and Th2 cell. J Exp Med 187:129 134

Bradding P, Feather IH, Howarth PH, Mueller R, Roberts JA, Britten K, Bews JPA, Hunt TC, Okayama Y, Heusser CH, Bullock GR, Church MK, Holgate ST (1992) Interleukin 4 is localized and released by human mast cells. J Exp Med 176:1381 1386

Bradley LM, Dalton DK, Croft M (1996) A direct role for IFN-γ in regulation of Th1 cell development. J Immunol 157:1350 1358

Briner TJ, Kuo MC, Keating KM, Rodgers BL, Greenstein JI (1993) Peripheral T-cell tolerance induced in naive and primed mice by subcutaneous injection of peptides from the major cat allergen Fel d1. Proc Natl Acad Sci USA 90:7608 7612

Brinkmann V, Kinzel B, Kristofic C (1996) TCR-independent activation of human CD4⁺ 45RO- T cells by anti-CD28 plus IL-2: induction of clonal expansion and priming for a Th2 phenotype. J Immunol 156:4100 4106

Brown DR, Fowell DJ, Corry DB, Wynn TA, Moskowitz NH, Cheever AW, Locksley RM, Reiner SL (1996) β2-microglobulin-dependent NK1.1[+] T cells are not essential for T helper cell two immune responses. J Exp Med 184:1295–1304

Bruce IN, Harland RW, McBride NA, MacMahon J (1993) Trends in the prevalence of asthma and dispnoea in first year university students 1972–89. Q J Med 86:425–430

Brunner T, Heusser CH, Dahinden CA (1993) Human peripheral blood basophils primed by interleukin 3 (IL-3) produce IL-4 in response to immunoglobulin E receptor stimulation. J Exp Med 177:605–611

Brusco A, Cariota U, Bottaro A, Boccazzi C, Plebani A, Ugazio AG, Galanello R, Guerra MG, Carbonara A (1995) Variability of the immunoglobulin heavy chain constant region locus: a population study. Hum Genet 95:319–326

Burastero SE, Crimi E, Balbo A, Vavassori M, Brogonovo B, Gaffi D, Frittoli E, Casorati G, Rossi GA (1995) Oligoclonality of lung T lymphocytes following exposure to allergen in asthma. J Immunol 155:5836–5846

Chan SC, Brown MA, Willcox TM, Li SH, Stevens SR, Tara D, Hanifin JM (1996) Abnormal IL-4 gene expression by atopic dermatitis T lymphocytes is reflected in altered nuclear protein interactions with IL-4 transcriptional regulatory element. J Invest Dermatol 106:1131–1136

Chehimi J, Trinchieri G (1995) Interleukin-12: a Bridge between innate resistance and adaptive immunity with a role in infection and acquired immunodeficiency. J Clin Immunol 14:149–161

Cho SS, Bacon CM, Sudarshan C, Rees RC, Finblom D, Pine R, O'Shea JJ (1996) Activation of STAT4 by IL-12 and IFN-alpha. Evidence for the involvement of ligand-induced tyrosine and serine phosphorylation. J Immunol 157:4781–4789

Chu RS, Targoni OS, Krieg AM, Lehmann PV, Harding CV (1997) CpG oligodeoxynucleotides act as adjuvants that switch on T helper 1 (th1) immunity. J Exp Med 186:1623–1631

Cocks BG, Chang CCJ, Carballido JM, Yssel H, de Vries JE, Aversa G (1995) A novel receptor involved in T cell activation. Nature 376:260–263

Conrad DH, Ben-Sasson S, Le Gros GG, Finkelman FD, Paul WE (1990) Infection with *Nippostrongylus brasiliensis* or injection of anti-IgD antibodies markedly enhances Fc-receptor-mediated interleukin 4 production by non-B, non-T cells. J Exp Med 171:1497–1508

Constant SL, Bottomly K (1997) Induction of TH1 and TH2 CD4[+] T cell responses: The alternative approaches. Annu Rev Immunol 15:297–322

Cookson WOCM (1998) Genetics and asthma. Res Immunol 149:181–187

Corne J, Djukanovic R, Thomas L, Warner J, Botta L, Grandordy B, Gygax D, Heusser C, Patalano F, Richardson W, Kilchherr E, Staehelin T, Davis F, Gordon W, Sun L, Liou R, Wang G, Chang TW, Holgate S (1997) The effect of intravenous administration of a chimeric anti-IgE antibody on serum IgE levels in atopic subjects: efficacy, safety, and pharmacokinetics. J Clin Invest 99(5):879–887

Coyle AJ, Tsuyuki S, Bertrand C, Huang S, Aguet M, Alkan SS, Anderson GP (1996) Mice lacking the IFN-gamma receptor have an impaired ability to resolve a lung eosinophilic inflammatory response associated with a prolonged capacity of T cells to exhibit a Th2 cytokine profile. J Immunol 156:2680–2685

Croft M, Swain SL (1995) Recently activated naive CD4 T cells can help resting B cells, and can produce sufficient autocrine IL-4 to drive differentiation to secretion of T helper 2-type cytokines. J Immunol 154:4269–4282

DeKruyff RH, Fang Y, Wolf SF, Umetsu DT (1995) IL-12 inhibits IL-4 synthesis in keyhole limpet hemocyanin-primed CD4[+] T cells through an effect on antigen-presenting cells. J Immunol 154:2578–2587

Del Prete GF, De Carli M, Mastromauro C, Macchia D, Biagiotti R, Ricci M, Romagnani S (1991) Purified protein derivative of Mycobacterium tuberculosis and excretory-secretory antigen(s) of Toxocara canis expand in vitro human T cells with stable and opposite (type 1 T helper or type 2 T helper) profile of cytokine production. J Clin Invest 88:346–351

Del Prete GF, De Carli M, D'Elios MM, Maestrelli P, Ricci M, Fabbri L, Romagnani S (1993a) Allergen exposure induces the activation of allergen-specific Th2 cells in the airway mucosa of patients with allergic respiratory disorders. Eur J Immunol 23:1445–1449

Del Prete GF, De Carli M, Almerigogna F, Giudizi M-G, Biagiotti R, Romagnani S (1993b) Human IL-10 is produced by both type 1 helper (Th1) and type 2 helper (Th2) T cell clones and inhibits their antigen-specific proliferation and cytokine production. J Immunol 150:1–8

Del Prete GF, De Carli M, Almerigogna F, Daniel KC, D'Elios MM, Zancuoghi D, Vinante E, Pizzolo G, Romagnani S (1995a) Preferential expression of CD30 by human CD4[+] T cells producing Th2-type cytokines. FASEB J 9:81–86

Del Prete GF, De Carli M, D'Elios MM, Daniel KC, Smith CA, Thomas E, Romagnani S (1995b) CD30-mediated signaling promotes the development of human Th2-like T cells. J Exp Med 182:1–7

Demeure CE, Yang LP, Byun DG, Ishihara H, Vezzio N, Delespesse G (1995) Human naive CD4 T cells produce interleukin-4 at priming and acquire a Th2 phenotype upon repetitive stimulations in neutral conditions. Eur J Immunol 25:2722–2725

de Vries JE (1996) Molecular and biological characteristics of interleukin 13. Chem. Immunol. 63:204–218

Dudler T, Cantarelli Machado D, Kolbe L, Annand RR, Rhodes N, Gelb MH, Koelsch E, Suter M, Helm BA (1995) A link between catalytic activity, IgE-independent mast cell activation, and allergenicity of bee venom phospholipase A_2. J Immunol 155:2605–2613

Egan RW, Athwahl D, Chou C-C, Emtage S, Jehn C-H, Kung TT, Mauser PJ, Murgolo NJ, Bodmer MW (1995) Inhibition of pulmonary eosinophilia and hyperreactivity by antibodies to interleukin 5. Int. Arch. Allergy Immunol. 107:321–322

Erb KJ, Holloway JW, Sobeck A, Moll H, Le Gros G (1998) Infection of mice with Mycobacterium bovis bacillus Calmette-Guérin suppresses allergen-induced airway eosinophilia. J Exp Med 187:561–569

Evavold BD, Williams SG, Hsu BL, Buus S, Allen PM (1992) Complete dissection of the Hb (64–76) determinant using T helper 1 and T helper 2 clones and T cell hybridomas. J Immunol 148:347–53

Fahy JV, Fleming HE, Wong HH, Liu JT, Su JQ, Fick RB Jr, Boushey HA (1997) The effect of an anti-IgE monoclonal antibody on the early and late-phase responses to allergen inhalation in asthmatic subjects. Am J Respir Crit Care Med 155:1828–1834

Fasler S, Averso G, Terr A, Thestrup-Pedersen K, de Vries JE, Yssel Y (1995) Peptide-induced anergy in allergen-specific human Th2 cells results in lack of cytokine production and B cell help for IgE synthesis: reversal by Il-2, not by IL-4 or IL-13. J Immunol 155:4199–4206

Fennelly GJ, Flynn JL, ter Meulen V, Liebert UG, Bloom BR (1995) Recombinant bacille Calmette-Guérin priming against measles. J Infect Dis 172:698–705

Finkelman FD, Pearce EJ, Urban JF Jr, Sher A (1991a) Regulation and biological function of helminth-induced cytokine responses. Immunoparasitol Today 12: A62–66

Finkelman FD, Svetic A, Gresser I, Snapper C, Holmes J, Trotta PP, Katona IM, Gause WC (1991b) Regulation by interferon of immunoglobulin isotype selection and lymphokine production in mice. J Exp Med 174:1179–1188

Fitch FW, McKisic MD, Lancki DW, Gajewski TF (1993) Differential regulation of murine T lymphocyte subsets. Annu Rev Immunol 11:29–48

Forsthuber T, Yip HC, Lehmann PV (1996) Induction of Th1 and Th2 immunity in neonatal mice. Science 271:1728–1730

Gajewski TF, Fitch FW (1989) Anti-proliferative effect of IFN-γ in immune regulation. I. IFN-γ inhibits the proliferation of Th2 but not Th1 murine helper T lymphocyte clones. J Immunol 140:4245–4252

Gajewski TF, Fitch FW (1990) Anti-proliferative effect of IFN-gamma in immune regulation. IV. Murine CTL clones produce IL-3 and GM-CSF, the activity of which is masked by the inhibitory action of secreted IFN-gamma. J Immunol 144:548–556

Galli G, Annunziato F, Mavilia C, Romagnani P, Cosmi L, Manetti R, Pupilli C, Maggi E, Romagnani S (1998) Enhanced HIV expression during Th2 oriented responses explained by the opposite regulatory effect of IL-4 and IFN-γ on fusin/CXCR4. Eur J Immunol 28:1–11

Gammon G, Sercarz E (1989) How some T cells escape tolerance induction. Nature 342:183–185

Gavett SH, O'Hearn DJ, Li X, Huang S-K, Finkelman FD, Wills-Karp M (1995) Interleukin 12 inhibits antigen-induced airway hyperresponsiveness, inflammation, and Th2 cytokine expression in mice. J Exp Med 182:1527–1536

Gieni RS, Yang X, HayGlass KT (1993) Allergen-specific modulation of cytokine synthesis patterns and IgE responses in vivo with chemically modified allergen. J Immunol 150:302–310

Gieni RS, Yang X, HayGlass KT (1996) Limiting dilution analysis of CD4 T-cell cytokine production in mice administered native versus polymerized ovalbumin: directed induction of T-helper type 1-like activation. Immunology 87:119–126

Gleich GJ, Kita H, Adolphson CR (1994) Eosinophils. In: Frank MM, Austen KF, Claman HN, Unanue ER (eds) Samter's immunological diseases. Little Brown, Boston, pp 205–245

Gorham JD, Guler ML, Steen RG, Mackey AJ, Daly MJ, Frederick K, Dietrich WF, Murphy KM (1996) Genetic mapping of a locus controlling development of Th1/Th2 type responses. Proc Natl Acad Sci USA 93:12467–12472

Guery JC, Galbiati F, Smiroldo S, Adorini L (1996) Selective development of T helper (Th) 2 cells induced by continous administration of low dose soluble proteins to normal and β2-microglobulin–deficient BALB/c mice. J Exp Med 183:485–497

Hamid Q, Azzawi M, Ying S, Moqbel R, Wardlaw AJ, Corrigan CJ, Bradley B, Durham SR, Collins JV, Jeffrey PK, Quint DJ, Kay AB (1991) Expression of mRNA for interleukin-5 in mucosal bronchial biopsies from asthma. J Clin Invest 87:1541 1546

HayGlass KT, Stefura W (1991) Antigen-specific modulation of murine IgE and IgG2a responses with glutaraldehyde-polymerized allergen is independent of MHC haplotype and IgH allotype. Immunology 73:24 30

Hershey GK, Friedrich MF, Esswein LA, Thomas ML, Chatila TA (1997) The association of atopy with a gain-of-function mutation in the alpha subunit of the interleukin-4 receptor. N Engl J Med 337:1720-1725

Hilkens CMU, Messer G, Tesselaar K, van Rietschoten AGI, Kapsenberg ML, Wierenga EA (1996) Lack of IL-12 signaling in human allergen-specific Th2 cells. J Immunol 157:4316 4321

Ho CI, Hodge MR, Rooney JW, Glimcher LH (1996) The proto-oncogene c-maf is responsible for tissue-specific expression of Interleukin-4. Cell 85:973 983

Hodge MR, Rooney JW, Glimcher LH (1995) The proximal promoter of the IL-4 gene is composed of multiple essential regulatory sites that bind at least two distinct factors. J Immunol 154:6397 6405

Hoyne GF, Askonas BA, Hetzel C, Thomas WR, Lamb JR (1996) Regulation of house dust mite responses by intranasally administered peptide: transient activation of CD4 [1] T cells precedes the development of tolerance in vivo. Int Immunol 8:335 342

Hsieh CS, Macatonia SE, Tripp CS, Wolf SF, O'Garra A, Murphy KM (1993) Development of TH1 CD4 [1] T cells through IL-12 produced by Leisteria-induced macrophages. Science 260:547 49

Hsu CH, Chua KY, Tao MH, Huang SK, Hsieh KH (1996) Inhibition of specific IgE response in vivo by allergen-gene transfer. Int Immunol 8:1405 1411

Janeway C, Carding S, Jones B, Murray J, Portoles P, Rasmussen R, Saizawa K, West J, Bottomly K (1988) CD4 [1] T cells: specificity and function. Immunol Rev 101:39 80

Jardieu P (1995) Anti-IgE therapy. Curr Opin Immunol 7:779 782

Jones AC, Miles EA, Warner JO, Colwell BM, Bryant TN, Warner JA (1996) Fetal peripheral blood mononuclear cell proliferative responses to mitogenic and allergenic stimuli during gestation. Pediatr Allergy Immunol 7:109 116

Jourdan P, Abbal C, Nora N, Hori T, Uchiyama T, Vendrell JP, Bousquet J, Taylor N, Pene J, Yssel H (1998) IL-4 induces functional cell-surface expression of CXCR4 on human T cells. J Immunol 160:153 157

Jutel M, Pichler WJ, Skrbic D, Urwyler A, Dahinden C, Muller UR (1995) Bee venom immunotherapy results in decrease of IL-4 and IL-5 and increase of IFN-γ secretion in specific allergen-stimulated T cell cultures. J Immunol 154:4187 4194

Kalinski P, Hilkens CMU, Wierenga EA, van der Pouw-Kraan TCTM, van Lier RAW, Bos JD, Kapsenberg ML, Snijdewint FGM (1995) Functional maturation of human naive T helper cells in the absence of accessory cells. Generation of IL-4-producing T helper cells does not require exogenous IL-4. J Immunol 154:3753 3760

Kaplan MH, Schindler U, Smiley ST, Grusby MJ (1996) STAT 6 is required for mediating responses to IL-4 and for development of Th2 cells. Immunity 4:313 319

Klinman DM, Yamschikov G, Ishigatsubo Y (1997) Contribution of CpG motifs to the immunogenicity of DNA vaccines. J Immunol 158:3635 3639

Ko LJ, Yamanoto M, Leonard MW, George KM, Ting P, Engel JD (1991) Murine and human T-lymphocyte GATA-3 factors mediate transcription through a cis-regulatory element within the human T cell receptor a gene enhancer. Mol Cell Biol 11:2778 2784

Kopf M, Le Gros G, Bachmann M, Lamers MC, Bluthmann H, Kohler G (1993) Disruption of the murine IL-4 gene blocks Th2 cytokine responses. Nature 362:245 248

Kramer U, Behrendt H, Dolgner R, Kainka-Stanicke E, Oberbarnscheidt J, Sidaoui H, Schlipkoter W (1991) Auswirkung der Umweltbelastung auf allergologische Parameter bei 6-jahrigen Kindern. Ergbnisse einer Pilotstudie im Rahmen der Luftreinhalteplane von Nordrhein-Westfalen. In Ring J (Ed) Epidemiologie allergischer Ekrankungen. Munich: MMV Medizin, pp 165 178

Krieg AM, Yi AK, Matson S, Waldschmidt TJ, Bishop GA, Teasdale R, Koretzky GA, Klinman DM (1995) CpG motifs in bacterial DNA trigger direct B-cell activation. Nature 374:546 549

Kumar V, Bhardwaj V, Soares L, Alexander J, Sette A, Sercarz E (1995) Major histocompatibility complex binding affinity of an antigenic determinant is crucial for the differential secretion of interleukin 4/5 or interferon by T cells. Proc Natl Acad Sci USA 92:9510 9514

Kung TT, Stelts DM, Zurcher JA, Adams GK, Egan RW, Kreutner W, Watnick AS, Jones H, Chapman RW (1995) Involvement of IL-5 in a murine model of allergic pulmonary inflammation: prophylactic and therapeutic effect of an anti-IL-5 antibody. Am J Respir Cell Mol. Biol. 13:360 365

Lack G, Renz H, Saloga J, Bradley K, Loader J, Leung DYM, Gelfand EW (1994) Nebulized but not parenteral IFN-γ decreases IgE production and normalizes airways function in a murine model of allergen sensitization. J Immunol 152:2546–2554

Lamb JR, Zanders ED, Sewell W, Crumpton MJ, Feldmann M, Owen MJ (1987) Antigen-specific T cell unresponsiveness in cloned helper T cells mediated via the CD2 or CD3/Ti receptor pathways. Eur J Immunol 17:1641–1644

Launois P, Ohteki T, Swihart K, Robson MacDonald H, Louis JA (1995) In susceptible mice, Leishmania major induce very rapid interleukin-4 production by CD4⁺ T cells which are NK1.1. Eur J Immunol 25:3298–3307

Le Gros A, Ben Sasson SZ, Paul WE (1993) Anti-IL-4 diminishes in vivo priming for antigen-specific IL-4 production by T cells. J Immunol 150:2112–2120

Li-Weber M, Eder A, Ktaf-Czepa H, Krammer PH (1992) T cell-specific negative regulation of transcription of the human cytokine IL-4. J Immunol 148:1913–1918

Lichtenstein LM (1993) Allergy and the immune system. Sci Am 269:117–124

Liew FY, Millott SM, Schmidt JA (1990) A repetitive peptide of Leishmania can activate T helper type 2 cells and enhance disease progression. J Exp Med 172:1359–1365

Lord CJM Lamb JR (1996) TH2 cells in allergic inflammation: a target of immunotherapy. Clin Exp Allergy 26:756–765

MacGlashan D, White JM, Huang S-K, Ono SJ, Schroeder JT, Lichtenstein LM (1994) Secretion of IL-4 from human basophils. The relationship between IL-4 mRNA and protein in resting and stimulated basophils. J Immunol 152:3006–3016

Maggi E, Biswas P, Del Prete GF, Parronchi P, Macchia D, Simonelli C, Emmi L, De Carli M, Tiri A, Ricci M, Romagnani S (1991) Accumulation of Th2-like helper T cells in the conjunctiva of patients with vernal conjunctivitis. J Immunol 146:1169–1174

Maggi E, Parronchi P, Manetti R, Simonelli C, Piccinni MP, Santoni-Rugiu F, De Carli M, Ricci M, Romagnani S (1992) Reciprocal regulatory role of IFN-γ and IL-4 on the in vitro development of human Th1 and Th2 cells. J Immunol 148:2142–2147

Magram J, Connaughton SE, Warrier RR, Carvajal DM, Wu C-Y, Ferrante J, Stewart C, Sarmiento U, Faherty DA, Gately MK (1996) IL-12-deficient mice are defective in IFN-γ production and type 1 cytokine responses. J Immunol 4:471–481

Manetti R, Parronchi P, Giudizi MG, Piccinni MP, Maggi E, Trinchieri G, Romagnani S (1993) Natural killer cell stimulatory factor (interleukin-12) induces T helper type 1 (Th1)-specific immune responses and inhibits the development of IL-4-producing Th cells. J Exp Med 177:1199–1204

Manetti R, Gerosa F, Giudizi MG, Biagiotti R, Parronchi P, Piccinni MP, Sampognaro S, Maggi E, Romagnani S, Trinchieri G (1994) Interleukin 12 induces stable priming for interferon γ (IFN-γ) production during differentiation of human T helper (Th) cells and transient IFN-γ production in established Th2 cell clones. J Exp Med 179:1273–1283

Manetti R, Annunziato F, Tomasevic L, Giannò V, Parronchi P, Romagnani S, Maggi E (1995) Poly-inosinic acid: polycytidylic acid promotes T helper type1-specific immune responses by stimulating macrophage production of IFN-α and interleukin-12. Eur J Immunol 25:2656–2660

Marinaro M, Staats HF, Hiroi T, Jackson RJ, Coste M, Boyaka PN, Okahashi N, Yamamoto M, Kiyono H, Bluethmann H, Fujihashi K, McGhee JR (1995) Mucosal adjuvant effect of cholera toxin in mice results from induction of T helper 2 (Th2) cells and IL-4. J Immunol 155:4621–4629

Marsh DG, Neely JD, Breazeale DR, Ghosh B, Freidhoff LR, Ehrlich-Kautzky E, Schou C, Krishnaswamy G, Beaty TH (1994) Linkage analysis of IL-4 and other chromosome 5q31.1 markers and total serum immunoglobulin E concentrations. Science 264:1152–1156

Matricardi PM, Rosmini F, Ferrigno L, Nisini R, Rapicetta M, Chionne P, Stroffolini T, Pasquini P, D'Amelio R (1997) Cross sectional retrospective study of prevalence of atopy among Italian military students with antibodies against hepatitis A virus. BMJ 314:999–1003

Matsuoka T, Kohrogi H, Ando M, Nishimura Y, Matsushita S (1996) Altered TCR ligands affect antigen-presenting cell responses. Up-regulation of IL-12 by an analogue peptide. J Immunol 157:4837–4843

McHugh SM, Deighton J, Stewart AG, Lachmann PJ, Ewan PW (1995) Bee venom immunotherapy induces a shift in cytokine responses from a Th2 to a Th1 dominant pattern: comparison of rush and conventional immunotherapy. Clin Exper Allergy 25:828–838

McMenamin C, Holt PG (1993) The natural immune response to inhaled soluble protein antigens involves major histocompatibility complex (MHC) class I-restricted CD8⁺ T cell-mediated but MHC class-II-restricted CD4⁺ T cell-dependent immune deviation resulting in selective suppression of immunoglobulin E production. J Exp Med 178:889–899

Mingari MC, Maggi E, Cambiaggi A, Annunziato F, Schiavetti F, Manetti R, Moretta L, Romagnani S (1996) In vitro development of human CD4⁺ thymocytes into functionally mature Th2 cells. Exogenous IL-12 is required for priming thymocytes to the production of both Th1 cytokines and IL-10. Eur J Immunol 26:1083 1087

Miyamoto T, Takafuji S (1991) Environment and allergy. In: Ring J, Przybilla B (eds) New trends in allergy III. Springer-Verlag, Berlin Heidelberg New York, pp 459 468

Mocci S, Coffman RL (1995) Induction of a Th2 population from a polarized Leishmania-specific Th1 population by in vitro culture with IL-4. J Immunol 154:3779 3787

Moqbel R, Ying S, Barkans J, Newman TM, Kimmitt P, Vakelin M, Taborda-Barata L, Meng Q, Corrigan CJ, Durham SR, Kay AB (1995) Identification of mRNA for interleukin-4 in human eosinophils with granule localization and release of the translated product. J Immunol 155:4939 4947

Mori A, Suko M, Kaminuma P, Nishizaki Y, Nagahori T, Mikami T, Ohmura T, Hosino A, Hasakura Y, Okudaira H (1996) Enhanced production and gene expression of IL-5 in bronchial asthma. Possible management of atopic diseases with IL-5 specific gene transcription inhibitor. In: Sehon A, Kraft D (eds) Molecular biology of allergens and the atopic immune response. Plenum, New York

Morris SC, Madden KB, Adamovicz JJ, Gause WC, Hubbard BR, Gately MK, Finkelman FD (1994) Effects of IL-12 on in vivo cytokine gene expression and Ig isotype selection. J Immunol 152:1047 1056

Mosmann TR, Cherwinski H, Bond MW, Giedli MA, Coffman RL (1986) Two types of murine T cell clone. I. Definition according to profiles of lymphokine activities and secreted proteins. J Immunol 136:2348 2357

Muller U, Fricker M, Carballido JM (1997) In: Ring J, Berendt H, (eds) New trends in allergy IV. Hamburg (in press)

Murphy E, Shibuya K, Hosken N, Openshaw P, Maino V, Davis K, Murphy K, O'Garra A (1996) Reversibility of T helper 1 and 2 populations is lost after long-term stimulation. J Exp Med 183:901 914

Murray JS, Kasselman JP, Schountz T (1995) High-density presentation of an immunodominant minimal peptide on B cells is MHC-linked to Th1-like immunity. Cell Immunol 66:9 15

Murray PJ, Aldovini A, Young RA (1996) Manipulation and potentiation of antimycobacterial immunity using recombinant bacille Calmette-Guérin strains that secrete cytokines. Proc Natl Acad Sci USA 93:934 939

Nabors GS, Afonso LCC, Farrell JP, Scott P (1995) Switch from a type 2 to a type 1 T helper cell response and cure of established Leishmania major infection in mice is induced by combined therapy with interleukin 12 and pentostam. Proc Natl Acad Sci USA 91:3142 3146

Nakajima H, Nakao A, Watanabe Y, Yoshida S, Iwamoto I (1994) IFN-α inhibits antigen-induced eosinophil and CD4⁺ T cell recruitment into tissue. J Immunol 153:1264 1270

Nakajima H, Gleich GJ, Kita H (1996) Constitutive production of IL-4 and IL-10 and stimulated production of IL-8 by normal peripheral blood eosinophils. J Immunol 156:4859 4866

Nonaka M, Nonaka R, Woolley K, Adelroth E, Miura K, Okhawara Y, Glibetic M, Nakano K, O'Byrne P, Dolovich J, Jordana M (1995) Distinct immunohistochemical localization of IL-4 in human inflamed airway tissues. J Immunol 155:3234 3244

Norman PS, Ohman JL, Long AA (1994) Clinical experience with T cell reactive peptides from cat allergen Feld 1. J Allergy Clin Immunol 93:231 235

Ochensberger B, Rihs S, Brunner T, Dahinden CA (1995) IgE-independent interleukin-4 expression and induction of a late phase of leukotriene C4 formation in human blood basophils. Blood 86:4039 4049

Okayama Y, Pette Frere C, Kassel O, Semper A, Quint D, Tunon-de-Lara MJ, Bradding P, Holgate ST, Church MK (1995) IgE-dependent expression of mRNA for IL-4 and IL-5 in human lung mast cells. J Immunol 155:1796 1808

Parronchi P, Macchia D, Piccinni M-P, Biswas P, Simonelli C, Maggi E, Ricci M, Ansari AA, Romagnani S (1991) Allergen- and bacterial antigen-specific T-cell clones established from atopic donors show a different profile of cytokine production. Proc Natl Acad Sci USA 88; 4538 4542

Parronchi P, De Carli M, Manetti R, Simonelli C, Piccinni MP, Macchia D, Maggi E, Del Prete G-F, Ricci M, Romagnani S (1992a) Aberrant interleukin (IL)-4 and IL-5 production in vitro by CD4⁺ helper T cells from atopic subjects. Eur J Immunol 22:1615 1620

Parronchi P, De Carli M, Manetti R, Simonelli C, Sampognaro S, Piccinni MP, Macchia D, Maggi E, Del Prete GF, Romagnani S (1992b) IL-4 and IFN(s) (alpha and gamma) exert opposite regulatory effects on the development of cytolytic potential by Th1 or Th2 human T cell clones. J Immunol 149:2977 2982

Parronchi P, Mohapatra S, Sampognaro S, Giannarini L, Wahn U, Chong P, Mohapatra SS, Maggi E, Renz H, Romagnani S (1996) Modulation by IFN-α of cytokine profile, T cell receptor repertoire and peptide reactivity of human allergen-specific T cells. Eur J Immunol 26:697 703

Parronchi P, Sampognaro S, Annunziato F, Brugnolo F, Radbruch A, Di Modugno F, Ruffilli A, Romagnani S, Maggi E (1998) Influence of both T cell receptor repertoire and severity of the atopic status on the cytokine secretion profile of Parietaria officinaris-specific T cells. Eur J Immunol 28: 37 46

Perez VL, Lederer JA, Lichtman AH, Abbas AK (1995) Stability of Th1 and Th2 populations. Int Immunol 7:869 875

Pfeiffer C, Stein J, Southwood S, Ketelaar H, Sette A, Bottomly K (1995) Altered peptide ligands can control CD4 T lymphocyte differentiation in vivo. J Exp Med 181:1569 1574

Piccinni MP, Macchia D, Parronchi P, Giudizi M-G, Bani D, Aterini R, Grossi A, Ricci M, Maggi E, Romagnani S (1991) Human bone marrow non-B, non-T cells produce interleukin-4 in response to cross-linkage of Fcε and Fcγ receptors. Proc Natl Acad Sci USA 88:8656 8660

Piccinni M-P, Giudizi M-G, Biagiotti R, Annunziato F, Manetti R, Giannarini L, Parronchi P, Sampognaro S, Romagnani S, Maggi E (1995) Progesterone favors the development of human T helper (Th) cells producing Th2-type cytokines and promotes both IL-4 production and membrane CD30 expression in established Th1 clones. J Immunol 155:128 133

Piccinni MP, Beloni L, Giannarini L, Livi C, Scarselli G, Romagnani S, Maggi E (1996) Abnormal production of Th2-type cytokines (IL-4 and IL-5) by T cells from newborns with atopic parents. Eur J Immunol 26:2293 2298

Piccinni MP, Beloni L, Livi C, Maggi E, Scarselli G, Romagnani S (1998) Role of type 2 T helper (Th2) cytokines and leukemia inhibitory factor (LIF) produced by decidual T cells in unexplained recurrent abortions. Nat Med (in press)

Plaut M, Pierce JH, Watson CJ, Hanley HJ, Nordan RP, Paul WE (1989) Mast cell lines produce lymphokines in response to cross-linkage of FcεRI or to calcium ionophores. Nature 339:64 67

Racioppi L, Ronchese F, Matis LA, Germain RN (1993) Peptide-major histocompatibility complex class II complexes with mixed agonist/antagonist properties provide evidence for ligand-related differences in T cell receptor-dependent intracellular signaling. J Exp Med 177:1047 1060

Raz E, Tighe E, Sato Y, Corr M, Dudler JA, Roman M, Swain SL, Spiegelberg HL, Carson DA. (1996) Preferential induction of a Th1 immune response and inhibition of specific IgE antibody formation by plasmid DNA immunization. Proc Natl Acad Sci USA 93:5141 5145

Renz H, Enssle K, Lauffer L, Kurrle R, Gelfand EW (1995) Inhibition of allergen-induced IgE and IgG1 production by soluble IL-4 receptor. Int Arch Allergy Immunol 106:46 54

Ring J, Behrendt H (1993) Allergy and IgE production: role of infection and environmental pollution. Allergy J 2:27 30

Robinson DS, Hamid Q, Ying S, Tsicopoulos A, Barkans J, Bentley AM, Corrigan CJ, Durham SR, Kay AB (1992) Predominant Th2-like bronchoalveolar T-lymphocyte population in atopic asthma. New Engl J Med 326:295 304

Rogge L, Barberis L, Passini N, Presky DH, Gubler U, Sinigaglia F (1997) Selective expression of an interleukin 12 receptor component by human T helper 1 cells. J Exp Med 185:825 831

Romagnani S (1991) Human Th1 and Th2: doubt no more. Immunol Today 12:256 257

Romagnani S (1992) Induction of T$_H$1 and T$_H$2 response: a key role for the 'natural' immune response? Immunol. Today 13:379 81

Romagnani S (1994a) Regulation of the development of type 2 T-helper cells in allergy. Curr Opin Immunol 6:838 846

Romagnani S (1994b) Lymphokine production by human T cells in disease states. Annu Rev Immunol 12:227 257

Romagnani S (1995a) Atopic allergy and other hypersensitivities. Editorial overview: technological advances and new insights into pathogenesis prelude novel therapeutic strategies. Curr Opin Immunol 7:745 750

Romagnani S (1995b) Biology of human Th1 and Th2 cells. J Clin Immunol 15:121 129

Romagnani S (1996) Th1 and Th2 in human diseases. Clin Immunol Immunopathol 80:225 235

Romagnani S (1997) The Th1/Th2 paradigm. Immunol Today 18:263 266

Romagnani S, Del Prete GF, Maggi E, Parronchi P, De Carli M, Manetti R, Piccinni MP, Almerigogna F, Giudizi MG, Biagiotti R, Sampognaro S (1993) Human Th1 and Th2 cells: regulation of development and role in protection and disease. In: J Gergely et al (eds) Progress in immunology VIII. Springer Hungarica, pp 239 246

Sallusto F, Mackay CR, Lanzavecchia A (1997) Selective expression of the eotaxin receptor CCR3 by human T helper 2 cells. Science 277:2005 2007

Schmitz J, Thiel A, Kuhn R, Rajewsky K, Muller W, Assenmacher M, Radbruch A (1994) Induction of interleukin 4 (IL-4) expression In T helper (Th) cells is not dependent on IL-4 from non-T cells. J Exp Med 179:1349 1353

Schultz-Larsen F (1993) The epidemiology of atopic dermatitis. Monogr Allergy 31:9 28

Secrist H, Chelen CJ, Wen Y, Marshall JD, Umetsu DT (1993) Allergen immunotherapy decreases interleukin 4 production in CD4⁺ T cells from allergic individuals. J Exp Med 178:2123 2130

Seder R, Paul WE, Davis MM, Fazekas de St. Groth B (1992a) The presence of interleukin 4 during in vitro priming determines the lymphokine-producing potential of CD4⁺ T cells from T cell receptor transgenic mice. J Exp Med 176:1091 1098

Seder RA, Paul WE, Davis MM, Fazekas de St. Groth B (1992b) The presence of interleukin-4 during in vivo priming determines the lymphokine-producing potential of CD4⁺ T cells from T cell receptor transgenic mice. J Exp Med 176:1091 1098

Seder RA, Gazzinelli R, Sher A, Paul WE (1993) Interleukin 12 acts directly on CD4⁺ T cells to enhance priming for interferon O production and diminishes interleukin 4 inhibition of such priming. Proc Natl Acad Sci USA 90:10188 10192

Shaheen SO, Aaby P, Hall AJ, Barker DJP, Heyes CB, Shiell AW, Goudiaby A (1996) Measles and atopy in Guinea-Bissau. Lancet 347:1792 1796

Shimoda K, van Deursen J, Sangster MY, Sarawar SR, Carson RT, Tripp RA, Chu C, Quelle FW, Nosaka T, Vognali DAA, Doherty PC, Grosveld G, Paul WE, Ihle JN (1996) Lack of IL-4-induced Th2 response and IgE class switching in mice with disrupted Stat6 gene. Nature 380:630 632

Shirikawa T, Enomoto T, Shimazu S, Hopkin JM (1997) The inverse association between tuberculin responses and atopic disorders. Science 275:77 79

Skeiky JAW, Gauderian JA, Benson DR, Bacelar O, Carvalho EM, Kubin M, Badaro R, Trinchieri G, Reed SG (1995) A recombinant Leishmania antigen that stimulates human peripheral blood mononuclear cells to express a Th1-type cytokine profile and to produce interleukin 12. J Exp Med 181:1527 1537

Soloway P, Fish S, Passmore H, Gefter M, Coffee R, Manser T (1991) Regulation of the immune response to peptide antigens: differential induction of immediate-type hypersensitivity and T cell proliferation due to changes in either peptide structure or major histocompatibility complex haplotype. J Exp Med 174:847 858

Sornasse T, Larenas PV, Davis KA, de Vries JE, Yssel H (1996) Differentiation and stability of T Helper 1 and 2 cells derived from naive human neonatal CD4⁺ T cells, analyzed at single-cell level. J Exp Med 184:473 483

Swain SL, Weinberg AD, English M, Huston G (1990) IL-4 directs the development of Th2-like helper effectors. J Immunol 145:3796 3806

Szabo SJ, Glimcher LH, Ho I-C (1997) Genes that regulate interleukin-4 expression in T cells. Curr Opin Immunol 9:776 781

Szabo SJ, Jacobson NG, Dighe AS, Gubler U, Murphy KM (1995) Developmental commitment to the Th2 lineage by extinction of IL-12 signaling. Immunity 2:666 675

Takeda K, Tanaka T, Shi W, Matsumoto M, Minami M, Kashiwamura SI, Nakanishi K, Yoshida N, Kishimoto T, Akira S (1996) Essential role of Stat6 in IL-4 signalling. Nature 380:627 630

Tang D, DeVit M, Johnston SA (1992) Genetic immunization is a simple method for eliciting an immune response. Nature 365:152 154

Thierfelder S, Mocikat R, Mysliwietz J, Lindhofer M, Kremmer E (1995) Immunosuppression by Fc region-mismatched anti-T cell antibody treatment. Eur J Immunol 25:2242 2246

Tsitoura DC, Verhoef A, Gelder CM, O'Heir R, Lamb JR (1996) Altered T cell ligands derived from a major house dust mite allergen enhance IFN-gamma but not IL-4 production by human CD4⁺ T cells. J Immunol 157:2160 2165

Urban JF Jr, Madden KB, Cheever AW, Trotta PP, Katona IM, Finkelman FD (1993) IFN inhibits inflammatory responses and protective immunity in mice infected with the nematode parasite, *Nippostrongylus brasiliensis*. J Immunol 151:7086 7094

Urban JF, Maliszewski CR, Madden KB, Katona IM, Finkelman FD (1995) IL-4 treatment can cure established gastrointestinal nematode infections in immunocompetent and immunodeficient mice. J Immunol 154:4675 4684

VanCott JL, Staats HF, Pascual DW, Roberts M, Chatfield SN, Yamamoto M, Coste M, Carter PB, Kiyono H, McGhee JR (1996) Regulation of mucosal and systemic antibody responses by T helper cell subsets, macrophages, and derived cytokines following oral immunization with live recombinant *Salmonella*. J Immunol 156:1504 1514

van der Heijden FL, Wierenga EA, Bos JD, Kapsenberg ML (1991) High frequency of IL-4-producing CD4⁺ allergen specific T lymphocytes in atopic dermatitis lesional skin. J Invest Dermatol 97:389 394

van der Weid T, Beebe AM, Roopenian DC, Coffman RL (1996) Early production of IL-4 and induction of Th2 responses in the lymph node originate from an MHC class I- independent CD4⁺ NK1.1⁻ T cell population. J Immunol 157:4421 4427

van't Hof W, Driedijk PC, van den Berg M, Beck-Singer AG, Jung G, Aalberse RC (1991) Epitope mapping of the Dermatophagoides pteronyssinus house dust mite major allergen Der p II using overlapping synthetic peptides. Mol Immunol 28:1225 1232

Varney VA, Hamid Q, Gaga M, Ying S, Jacobson M, Frew AJ, Kay AB, Durham SR (1993) Influence of grass pollen immunotherapy on cellular infiltration and cytokine mRNA expression during allergen-induced late-phase cutaneous responses. J Clin Invest 92:644 651

Vicari AP, Mocci S, Openshaw P, O'Garra A, Zlotnik A (1996) Mouse γδ NK1.1⁺ thymocytes specifically produce interleukin-4, are major hystocompatibility complex Class I independent, and are developmentally related to αβ TCR⁺ NK1.1⁺ thymocytes. Eur J Immunol 26 1424:1429

Villacres-Eriksson M (1995) Antigen presentation by naive macrophages, dendritic cells and B cells to primed T lymphocytes and their cytokine production following exposure to immunostimulating complexes. Clin Exp Immunol 102:46 52

Wang Z-E, Zheng S, Corry DB, Dalton DK, Seder RA, Reiner SL Locksley RM (1994) Interferon γ-independent effects of interleukin 12 administered during acute or established infection due to *Leishmania major*. Proc Natl Acad Sci USA 91:12932 12936

Wegmann TG, Lin H, Guilbert L, Mosmann RT (1993) Bidirectional cytokine interactions in the maternal-fetal relationship: is successful pregnancy a Th2 phenomenon? Immunol Today 14:353 356

Wenner C, Guler ML, Macatonia SE, O'Garra A, Murphy KM (1996) Roles of IFN-γ and IFN-α in IL-12-induced Th1 development. J Immunol 156:1442 1447

Wershil BK, Theodos CM, Galli SJ, Titus RG (1994) Mast cells augment lesion size and persistence during experimental *Leishmania major* infection in the mouse. J Immunol 152:4563 4571

Wierenga EA, Snoek M, de Groot C, Chretien I, Bos JD, Jansen HM, Kapsenberg ML (1990) Evidence for compartmentalization of functional subsets of CD4⁺ T lymphocytes in atopic patients. J Immunol 144:4651 4656

Wynn TA, Jankovic D, Hieny S, Zioncheck K, Jardieu P, Cheever AW, Sher A (1995) IL-12 exacerbates rather than suppresses T helper 2-dependent pathology in the absence of endogenous IFN-γ. J Immunol 154:3999 4009

Xu-Armano J, Kiyono H, Jackson RJ, Staats HF, Fujiashi K, Burrows PD, Elson CO, Pillai S, McGhee JR (1993) Helper T cell subsets for immunoglobulin A responses: oral immunization with tetanus toxoid and cholera toxin as adjuvant selectively induces Th2 cells in mucosa associated tissues. J Exp Med 178:1309 1320

Yang L-P, Demeure D-G, Vezzio CE, Delespesse G (1995) Default development of cloned human naive CD4 T cells into interleukin-4- and interleukin-5-producing effector cells. Eur J Immunol 12:3517 3520

Yashimoto T, Paul WE (1994) CD4ᵖᵒˢ, NK1.1ᵖᵒˢ T cells promptly produce interleukin 4 in response to in vivo challenge with anti-CD3. J Exp Med 179:1285 1295

Yashimoto T, Bendelac A, Watson C, Hu-Li J, Paul WE (1995a) CD-1-specific, NK1.1ᵖᵒˢ T cells play a key in vivo role in a Th2 response and in IgE production. Science 270:1845 1847

Yashimoto T, Bendelac, A, Hu-Li J, Paul WE (1995b) Defective IgE production by SJL mice is linked to the absence of a subset of T cells that promptly produce IL-4. Proc Natl Acad Sci USA 92:11931 11934

Zheng W, Flavell RA (1997) The transcription factor GATA-3 is necessary and sufficient for Th2 cytokine gene expression inCD4 T cells. Cell 89:587 596

Zingoni A, Soto H, Hedrick JA, Stoppacciaro A, Storlazzi CT, Sinigaglia F, D'Ambrosio D, O'Garra A, Robinson D, Rocchi M, Santoni A, Zlotnik A, Napolitano M (1998) The chemokine receptor CCR8 is preferentially expressed in Th2 but not Th1 cells. J Immunol 161:547 551

Interleukin-12: Basic Principles and Clinical Applications

G. Trinchieri[1] and P. Scott[2]

1   Introduction 57

2   The Interleukin-12 Molecule: Its Genes and Its Receptor 58

3   Cell Types Producing Interleukin-12 60

4   Molecular Control of Interleukin-12 Production 61

5   In Vitro Activities of Interleukin-12 62

6   Interleukin-12 Induction of Th1 Responses 63

7   Role of Interleukin-12 in Infections 64

8   Use of Interleukin-12 as an Adjuvant for Infectious Diseases 64

9   Use of Interleukin-12 in Tumor Therapy 67

10 Use of Interleukin-12 as an Immunotherapeutic Agent 69

11 Interleukin-12: Future Use in Clinical Settings 71

References 72

1 Introduction

The most important advance in the last 10 years in our understanding of how to direct the immune response by vaccination or immunotherapy has been the description and subsequent refinement of the T helper type1/2 (Th1/2) paradigm (Mosmann and Coffman 1989). This paradigm has provided the framework necessary to formulate basic questions related to defining the cues pathogens provide that shape the immune response. A major advance in this area came with the discovery of interleukin-12 (IL-12) (Kobayashi et al. 1989; Stern et al. 1990) and the subsequent demonstration that IL-12 promotes the development of Th1 cells in vitro (Hsieh et al. 1993; Manetti et al. 1993). Thus, in naive T cell populations exposure to antigen in the presence of IL-12 for several days, followed by

[1]The Wistar Institute, 3601 Spruce Street, Philadelphia, PA 19104, USA
[2]Department of Pathobiology, University of Pennsylvania, School of Veterinary Medicine, 3800 Spruce Street, Philadelphia, PA 19104-6008, USA

restimulation with antigen alone, led to the development of interferon (IFN)-γ producing T cells. Moreover, it was found that one could link together IL-12, the innate immune response, pathogens, and Th1 cell development. This was done by showing that bacteria, such as *Listeria monocytogenes*, induced Th1 cell development and that this occurred by stimulation of macrophages to produce IL-12 (D'ANDREA et al. 1992; HSIEH et al. 1993). This observation has led to the description of a common pathway leading from the innate immune response to adaptive immunity, in which intracellular pathogens stimulate macrophages to produce IL-12, which promotes the development of Th1 cells from a naive cell population. This pathway can now be exploited to develop approaches for the design of new immunotherapies and vaccines.

2 The Interleukin-12 Molecule: Its Genes and Its Receptor

Interleukin-12 is a heterodimeric cytokine, composed of a heavy chain of 40 kDa (p40) and a light chain of 35 kDa (p35), originally described with the names of natural killer stimulatory factor (NKSF) (KOBAYASHI et al. 1989) or cytotoxic lymphocyte maturation factor (CLMF) (STERN et al. 1990). IL-12 is produced within a few hours of infection, particularly in the case of bacteria and intracellular parasites, and acts as a proinflammatory cytokine, activating natural killer (NK) cells and, through its ability to induce IFN-γ production, enhancing the phagocytic and bacteriocidal activity of phagocytic cells and their ability to release proinflammatory cytokines, including IL-12 itself (D'ANDREA et al. 1992; KUBIN et al. 1994a). Furthermore, IL-12 produced during the early phases of infection and inflammation sets the stage for the ensuing antigen-specific immune response, favoring differentiation and function of Th1 T cells while inhibiting the differentiation of Th2 T cells (MANETTI et al. 1993, HSIEH et al. 1993). Thus, IL-12 in addition to being a potent proinflammatory cytokine, is a key immunoregulator molecule in Th1 responses. The two genes encoding the two chains of IL-12 are separated and unrelated; the gene encoding the p35 light chain has limited homology with other single chain cytokines, whereas the gene encoding the p40 heavy chain is homologous to the extracellular domain of genes of the hematopoietic cytokine receptor family. The p35 and the p40 chains are covalently linked to form a biologically active heterodimer (p70) (WOLF et al. 1991; GEARING and COSMAN 1991).

Two or more IL-12 binding affinities are observed on IL-12 responsive cells, and the receptors with the highest affinity, in the picomolar range, are probably responsible for IL-12 biological activity. Two chains of the IL-12 receptor, IL-12Rβ1 and IL-12 Rβ2, have been cloned and are members of the cytokine receptor superfamily and within that family are most closely related to gp130 (CHUA et al. 1994, 1995; PRESKY et al. 1996). Both IL-12Rβ1 and IL-12Rβ2 have a cytoplasmic region that contains the characteristic box I and II motifs found in other cytokine receptors. However, conserved cytoplasmic tyrosine residues are missing from the β1 subunit, whereas the IL-12 Rβ2 subunit contains three cytoplasmic tyrosine residues which are likely involved in signaling processes. Each subunit alone binds IL-12 with only low affinity ($K_d = 2–5$ nM). Coexpression of both receptor subunits results in both high affinity ($K_d = 50$ pM) and low affinity ($K_d = 5$nM) IL-12 binding sites (PRESKY et al. 1996). IL-12 p40 interacts mostly with IL-12Rβ1, whereas IL-12 p35 or possibly an epitope on IL-12 composed of both ligand subunits appears to interact mostly with the IL-12Rβ2 of the receptor complex (PRESKY et al. 1996).

The cells producing IL-12 secrete a large excess of the free p40 chain over the biologically active heterodimer, from a few-fold, as observed in activated phagocytic cells, to up to 100- to 1000-fold (D'ANDREA et al. 1992). Recombinant p40 IL-12, both human and murine, is secreted by transfected cells both as a disulfide-bond homodimer or as a monomer, and, in the mouse, homodimer p40 production has also been observed in vivo, although such observations have not been made in humans. Murine p40 homodimers bind to the IL-12Rβ1 chain with an affinity similar to that of the heterodimers and compete with the heterodimers for binding to the IL-12 receptor, effectively blocking the biological functions of IL-12 on murine cells (GILLESSEN et al. 1995; MATTNER et al. 1993). On human cells, the homodimers bind to the IL-12R with a much lower affinity than the heterodimers and act as antagonists only at much higher concentrations than on murine cells (LING et al. 1995). Thus, in the mouse, but not likely in humans, the IL-12 p40 homodimer may represent a physiologic antagonist of IL-12.

Resting T and NK cells do not express or express only at very low levels the IL-12R (DESAI et al. 1992); however, resting peripheral blood T and NK cells rapidly respond to IL-12 with IFN-γ production and enhancement of cytotoxic functions, suggesting that the receptor is present at least in a proportion to the cells and/or it can be rapidly activated in culture (KOBAYASHI et al. 1989; CHAN et al. 1991). Activation of T and NK cells induces up-regulation of IL-12R, as identified by low and high affinity binding and up-regulation at least of the IL-12Rβ1 gene (WU et al. 1997; GOLLOB et al. 1997). It should be noted that certain cell types, e.g., human B lymphoblastoid cell lines or normal B cells, also express the IL-12Rβ1 mRNA, without, in most cases, expressing IL-12 binding sites (BENJAMIN et al. 1996), suggesting that IL-12Rβ1 is essential but not sufficient for expression of functional, high affinity IL-12R and that the IL-12Rβ2 subunit may be more restricted in its expression than IL-12Rβ1.

Signal transduction through the IL-12R induces tyrosine phosphorylation of the Janus family kinases JAK2 and TYK2 and of the transcription factor STAT3 and STAT4 (JACOBSON et al. 1995; BACON et al. 1995a,b); IL-12 is the only inducer known to activate STAT4. Phosphorylation and activation of the 44 kDa MAP kinase may be responsible for a serine phosphorylation also observed in STAT4 upon stimulation of T cells within IL-12 (PIGNATA et al. 1994; BACON et al. 1995b). During developmental commitment of BALB/c CD4[+] T cells to the Th2 lineage, the ability of T cells to signal in response to IL-12 is extinct due to down-regulation of IL-12Rβ2 expression (SZABO et al. 1997).

3 Cell Types Producing Interleukin-12

Interleukin-12 was originally discovered as a product of Epstein-Barr virus (EBV)-transformed B cell lines, which all constitutively produce at least low levels of IL-12 p40 (KOBAYASHI et al. 1989; BENJAMIN et al. 1996). Although malignant or EBV-transformed B cell lines produce IL-12, the physiological relevance of IL-12 production from normal B cells remains to be established and subsequent studies suggested that phagocytic cells are the major physiological producers of IL-12 (D'ANDREA et al. 1992), a conclusion now supported by many in vitro and in vivo studies. Monocytes produce high levels of IL-12 p40 and p70 when stimulated by bacteria, such as heat-fixed *Staphylococcus aureus* or *Streptococcus* extracts, or bacterial products such as lipopolysaccharide (LPS) (D'ANDREA et al. 1992). In addition to monocytes, polymorphonuclear leukocytes (PMNs) also respond to LPS stimulation with production of IL-12 (CASSATELLA et al. 1995).

On both monocytes and PMNs, IFN-γ has a powerful enhancing effect on IL-12 production, probably potentiating it within inflammatory tissues (KUBIN et al. 1994a; CASSATELLA et al. 1995; MA et al. 1996). The ability of IFN-γ to enhance the production of IL-12 by phagocytic cells is of particular interest because IL-12 is a potent inducer of IFN-γ production by T and NK cells (KOBAYASHI et al. 1989; CHAN et al. 1991). Thus, IL-12-induced IFN-γ acts as a positive feedback mechanism in inflammation by enhancing IL-12 production. Also, because IL-12 is the major cytokine responsible for the differentiation of Th1 cells, which are producers of IFN-γ, the enhancing effect of IFN-γ on IL-12 production may represent a mechanism by which Th1 responses are maintained in vivo. The ability of IFN-γ to enhance IL-12 production is particularly evident and required for IL-12 production in the case of certain infectious agents, e.g., mycobacteria, which are rather poor inducers of IL-12 production (FLESCH et al. 1995). However, with many other inducers, such as LPS, *Toxoplasma gondii*, and *S. aureus*, IL-12 production in vivo and in vitro both precedes and is required for IFN-γ production (D'ANDREA et al. 1992; WYSOCKA et al. 1995; SCHARTON-KERSTEN et al. 1996).

The positive feedback amplification of IL-12 production mediated by IFN-γ represents a potentially dangerous mechanism leading to uncontrolled cytokine production. There are, however, potent mechanisms of down-regulation of IL-12 production and of the ability of T and NK cells to respond to IL-12. IL-10 is a potent inhibitor of IL-12 production by phagocytic cells; the ability of IL-10 to suppress production of IFN-γ and other Th1 cytokines is primarily due to its inhibition of IL-12 production from antigen-presenting cells (APCs) as well as by inhibition of expression of costimulatory surface molecules (e.g., B7) and soluble cytokines, e.g., tumor necrosis factor (TNF)-α, IL-1β (D'ANDREA et al. 1993). Another powerful inhibitor of IL-12 production is transforming growth factor (TGF)-β (D'ANDREA et al. 1995). IL-4 and IL-13 also partially inhibit IL-12 production (D'ANDREA et al. 1995), suggesting that Th2 cells, by producing cytokines such as IL-10, IL-4, and IL-13, suppress IL-12 production and prevent the emergence of a Th1 response.

Evidence that dendritic cells are producers of IL-12 came from studies demonstrating that, when endogenous IL-4 production is blocked, these cells induce a Th1 response which is prevented by neutralizing anti-IL-12 antibodies (MACATONIA et al. 1995). Extensive studies with both human and mouse dendritic cells and skin Langerhans cells have now confirmed that dendritic cells are efficient producers of the IL-12 that acts in inducing Th1 responses upon antigen presentation by these APCs.

In addition to the induction of IL-12 observed in response to infectious agents, activated T cells stimulate production of IL-12 by macrophages and dendritic cells (GERMANN et al. 1993; MACATONIA et al. 1995). The mechanism of this T cell-dependent induction of IL-12 is based on the interaction of CD40 ligand (CD40L) on the surface of activated T cells with CD40 on the APCs and can be mimicked by cross-linking CD40 on the surface of IL-12 producing cells with anti-CD40 antibodies or recombinant CD40L (SHU et al. 1995; CELLA et al. 1996). The induction of IL-12 by bacterial or other infectious agents and by activated T cells represents two independent pathways of APC activation, as clearly shown by the observation that spleen cells from CD40 KO mice are completely unable to produce IL-12 in response to activated T cells, but produce normal levels of IL-12 in response to endotoxin or *S. aureus* (MARUO et al. 1997). However, it is probable that during an infection or an immune response in vivo both pathways are activated, the T cell-independent one during the inflammatory phase of innate resistance and the T cell-dependent one during the subsequent adaptive immune response. Thus, the inflammatory pathway may be responsible for the initiation of the Th1 response and the T cell-dependent pathway for its maintenance.

4 Molecular Control of Interleukin-12 Production

Upon activation of phagocytic cells with LPS or *S. aureus*, accumulation of IL-12 p40 mRNA is observed within 2–4 h, slightly delayed compared to other pro-inflammatory cytokines such as TNF-α, and then subsides after several hours. The induction of p40 expression is largely controlled at the transcriptional level and both the enhancing effect of IFN-γ and the inhibitory effect of IL-10 are reflected in changes in the rate of IL-12 gene transcription (MA et al. 1996, 1997a). The promoter of the p40 gene is constitutively active in EBV-transformed cell lines and inducible in myeloid cell lines, but not in T cell lines; IFN-γ priming of the myeloid cells greatly enhances the activation of the promoter by LPS (MA et al. 1996). A region responsible for promoter induction and activity is between nucleotides −196 and −224 and it binds a series of IFN-γ- and LPS-induced nuclear proteins, including Ets2 and/or Ets-related factors (MA et al. 1997b). Promoter constructs with deletion or mutations in this region display a reduced but still detectable IFN-γ/ LPS inducible promoter activity, contributed in part by a site between −123 and −99 to which NF-kB binds (MURPHY et al. 1995).

Expression of the p35 gene is also up-regulated upon activation of phagocytic cells, although its ubiquitous constitutive expression has complicated analysis of its expression using non-purified cell preparations. p35 up-regulation is inhibited by IL-10, whereas IFN-γ enhances transcription and mRNA accumulation of the p35 gene (MA et al. 1996; D'ANDREA et al. 1993). In activated phagocytic cells and in B cell lines, p40 mRNA is approximately tenfold more abundant than p35 mRNA, explaining the overproduction and secretion of the free p40 chain over the p35-containing heterodimer.

5 In Vitro Activities of Interleukin-12

Interleukin-12 synergizes with other hematopoietic factors in enhancing survival and proliferation of early multipotent hematopoietic progenitor cells and lineage-committed precursor cells (JACOBSEN 1995). Although in vitro IL-12 has prevalently stimulatory effects on hematopoiesis, in vivo IL-12 treatment results in decreased bone marrow hematopoiesis and both transient anemia and neutropenia. The toxic effects of IL-12 on hematopoiesis are mostly mediated by IFN-γ and, in its absence, treatment with IL-12 results in stimulation of hematopoiesis only (ENG et al. 1995).

IL-12 induces T and NK cells to produce several cytokines and particularly IFN-γ (CHAN et al. 1991). IL-12-induced IFN-γ production requires the presence of low levels of both TNF and IL-1 (D'ANDREA et al. 1993). The importance of IL-12 as an IFN-γ inducer rests not only in its high efficiency at low concentrations, but also in its synergy with many other activating stimuli (CHAN et al. 1991). IL-12 is required for optimal IFN-γ production in vivo during immune responses, especially in bacterial or parasitic infections. In response to macrophage-produced IL-12, NK cells readily produce IFN-γ which activates macrophages and enhances their bacteriocidal activity, providing a mechanism of T cell-independent macrophage activation during the early phases of innate resistance.

IL-12 does not induce proliferation of resting peripheral blood T cells or NK cells, although it potentiates the proliferation of T cells induced by various mitogens and has a direct proliferative effect on preactivated T and NK cells (KOBAYASHI et al. 1989; GATELY et al. 1991; PERUSSIA et al. 1992). IL-12 is effective at lower concentrations than IL-2, although the levels of proliferation obtained with IL-12 are much lower than those observed with IL-2. However, costimulation through the CD28 receptor by the ligand B7 or anti-CD28 antibodies strongly synergizes with IL-12 in inducing both efficient T cell proliferation and cytokine production (KUBIN et al. 1994b; MURPHY et al. 1994). Because B7 is a surface molecule and IL-12 is a secreted product of APCs, their synergistic effect on T cells plays an important role in inducing T cell proliferation and IFN-γ production upon antigen presentation to T cells.

IL-12 also enhances the generation of cytotoxic T cells (CTLs) and lymphokine activated killer (LAK) cells and potentiates the cytotoxic activity of CTLs and NK cells (KOBAYASHI et al. 1989; GATELY et al. 1992). Some of the effects of IL-12 on cell-mediated cytotoxicity are due to increased formation of cytoplasmic granules and induction of transcription of genes encoding cytotoxic granule-associated molecules such as perforin and granzymes (SALCEDO et al. 1993; ASTE-AMEZAGA et al. 1994). The ability of IL-12 to induce expression of adhesion molecules on T and NK cells also may affect their cytotoxic activity and their ability to migrate to tissues (RABINOWICH et al. 1993).

6 Interleukin-12 Induction of Th1 Responses

Interleukin-12 is required for Th1 cell development during the immune response to pathogens, and the type of Th cell differentiation is most likely determined early after infection by the balance between IL-12 and IL-4 which favor Th1 and Th2 development, respectively. IL-12 is produced by phagocytic cells, other APCs, and possibly B cells, whereas IL-4 is produced by subsets of T cells and by mast cells.

The defining characteristic of the Th1 and Th2 cells is that they stably express the ability to produce certain cytokines but not others. IL-12 is particularly powerful, when present at the early time of clonal expansion, to prime T cells, both CD4$^+$ and CD8$^+$, for the ability to produce high levels of IFN-γ upon restimulation (MANETTI et al. 1994; TRINCHIERI et al. 1996). IL-12, produced during the inflammatory phase of infections or immune responses, induces NK cells and T cells to produce IFN-γ; then, IL-12, in cooperation with IFN-γ, induces the T cell clones expanding in response to the specific antigens to differentiate into Th1 cells by priming them for expression of cytokines such as IFN-γ and by exerting other positive or negative selective mechanisms, including, for example, deletion of IL-4 producing cells or preferential expansion of cells with a Th1-like phenotype. Once a Th1 response is induced in vivo, IL-12 is in most cases not necessary for maintaining it (GAZZINELLI et al. 1994). However, differentiated Th1 cells maintain IL-12 responsiveness and IL-12, produced by APCs during cognate antigen presentation to T cells, appears to be important, at least in autoimmune diseases, for optimal proliferation and cytokine production of the Th1 cells in response to antigens (SEDER et al. 1996).

Many of the effects of IL-12 on B cell activation and immunoglobulin isotype production could be interpreted as mediated by either a subset of Th cells or by their products, IFN-γ in particular (MORRIS et al. 1994). However, evidence has been provided that IL-12 may directly affect B cell proliferation and differentiation (JELINEK and BRAATEN 1995).

7 Role of Interleukin-12 in Infections

The proinflammatory functions of IL-12, its ability to stimulate innate resistance and to generate a Th1-type immune response, are essential for the resistance to different types of infection, particularly bacteria, fungi, and intracellular parasites. The most acute instance of IL-12 production resulting in IFN-γ induction is observed in the models of endotoxic-induced shock (OZMEN et al. 1994; HEINZEL et al. 1994; WYSOCKA et al. 1995). Similar pathogenetic mechanisms mediated by IL-12 are involved in the toxic shock-like syndromes induced by superantigenic exotoxins produced by gram-positive bacteria, e.g., *Streptococcus pyogenes* and *S. aureus* (LEUNG et al. 1995). The role of endogenous IL-12 in resistance to infection has been analyzed in many studies. Unlike what is observed in bacterial and intracellular parasite infection, IL-12 has a relatively minor role in the resistance to virus infection, and IL-12-independent mechanisms of IFN-γ production are operative in virus infection (BIRON and GAZZINELLI 1995).

8 Use of Interleukin-12 as an Adjuvant for Infectious Diseases

Experimental murine leishmanial infections provided one of the first examples of the importance of Th1 and Th2 cell subsets in disease. Because of the clear differential development of Th1 and Th2 cells in mice infected with *L. major*, this model has been used to define the factors that control the development of CD4+ T cell subsets and regulate those subsets once they have developed (LOCKSLEY and SCOTT 1991; SCOTT 1996, REINER and LOCKSLEY 1995). The study of the innate immune response to *L. major* in the resistant C3H mouse suggested that IL-12 might be useful in a vaccine against leishmaniasis. To test this possibility, BALB/c mice were immunized with a soluble leishmanial antigen in the presence or absence of IL-12 (AFONSO et al. 1994). The route of immunization was subcutaneous, followed 10 days later by a boost given intradermally. The immunized mice were then challenged with *L. major* and the course of infection monitored. In contrast to the controls, mice immunized with leishmanial antigen and IL-12 were protected against a normally fatal infection. These studies were the first to establish the efficacy of using IL-12 in a vaccine requiring cell-mediated immunity. More recently, it was shown that a single antigen and IL-12 could stimulate protective immunity (MOUGNEAU et al. 1995). Thus, a molecule termed LACK was administered with IL-12 to BALB/c mice, in the same protocol as described above, and the animals were protected against challenge infection. These results suggest that a single antigen given with IL-12 will be sufficient to induce a protective immune response in mice. This is an important observation since it suggests that no additional adjuvant is required other than a recombinant protein and IL-12 to induce protection. It should be pointed out, however, that this antigen may be somewhat unique, since the type of initial immune response that develops to this single dominant leishmanial antigen – whether anti-LACK T cells produce IL-4 or IFN-γ – may be critical in determining resistance and susceptibility to leishmaniasis (JULIA et al. 1996; REINER et al. 1993).

Our studies in C3H mice found that NK cells played a role in the early development of Th1 cells following infection with *L. major* (SCHARTON and SCOTT 1993; SCHARTON-KERSTEN and SCOTT 1995). Since IL-12 is associated with NK cell activation, we investigated what role NK cells might be playing in the development of vaccine-induced immunity. We found that administration of soluble leishmanial antigen and IL-12 induced an NK cell IFN-γ response in the lymph nodes draining the site of immunization. In contrast, by 2 weeks after immunization the NK cell response had diminished and was replaced by a CD4+ Th1 type response (AFONSO et al. 1994). This NK cell response was required for Th1 cell development following immunization with IL-12, since depletion of NK cells abrogated the development of a Th1 response in immunized mice (AFONSO et al. 1994). Similarly, it was found in a *Schistosoma mansoni* vaccine that the ability of IL-12 to promote Th1 cells was dependent upon the presence of an NK1.1+ cell population (MOUNTFORD et al. 1996). Our current interpretation of these results is that NK cells produce IFN-γ, which promotes the production of endogenous IL-12 and thus enhances the development of a Th1 response and at the same time inhibits Th2 cell development. This hypothesis is supported by our finding that depletion of IFN-γ at the time of immunization also abrogates Th1 cell development (AFONSO et al. 1994).

IL-12 has also been shown to be an effective adjuvant in several other systems. Significant protection against bacteria and helminths has been observed in mouse models when IL-12 has been part of the immunization protocol. For example, it was found that nonviable *Listeria monocytogenes* administered with IL-12 elicited protective immunity against a lethal challenge infection (MILLER et al. 1995). Notably, the protection observed following immunization with *Listeria* antigen and IL-12 was equivalent to that induced by sublethal infection with *Listeria*, which was the first demonstration of protection against *Listeria* in a vaccine using a nonviable organism. Similarly, the acellular *Bordetella pertussis* vaccine was improved significantly when IL-12 was used as an adjuvant (MAHON et al. 1996). Thus, with IL-12 the efficacy of the *Bordetella* subunit vaccine was equivalent to that observed with the whole cell vaccine (MAHON et al. 1996).

In schistosomiasis, the efficacy of vaccines in a mouse model using either attenuated organisms (irradiated cercariae) or antigen extracts was increased significantly when given with IL-12 (MOUNTFORD et al. 1996; WYNN et al. 1995b). In both cases, protection was associated with enhanced Th1 type responses, although the effects on antibody isotype varied. Mice immunized with attenuated cercariae exhibited enhanced levels of IgG1, which were unaffected by the presence of IL-12 (WYNN et al. 1995b). In contrast, mice immunized with soluble extracts of the worms, in combination with IL-12, exhibited a dramatic reduction in IgG1 levels (MOUNTFORD et al. 1996). In both cases, there was a significant reduction in the levels of total IgE associated with the infection.

Although endogenous IL-12 plays a minor role in the resistance to most viral infections, IL-12 was found to be an effective adjuvant also for viral vaccines. Thus, administration of IL-12 enhanced type-1 immune responses against respiratory syncytial virus and pseudorabies virus (SCHIJNS et al. 1995; TANG and GRAHAM 1995). In the pseudorabies system, the critical role that IFN-γ plays was directly demonstrated by showing that no protection was obtained with IL-12 in mice lacking the IFN-γ receptor (SCHIJNS et al. 1995). Studies with class I-restricted peptides strongly suggest that peptides and IL-12 alone may be sufficient to generate protective CTL activity, since mice immunized with influenza NP peptides and IL-12 were shown to be protected against influenza challenge infection (O'TOOLE et al. 1996).

Studies done to examine the important and practical issues of how best to administer IL-12 have shown that IL-12 can be effective as an adjuvant when administered in combination with antigen by the subcutaneous route, or when administered systemically separate from the antigen (BLISS et al. 1996a). With the model antigen TNP-KLH, the nature of the memory immune response induced when IL-12 is used as an adjuvant was also investigated. Immunization of BALB/c mice with TNP-KLH normally invokes a Th2 response, while systemic administration of IL-12 – at days −1, 0 and +1, relative to subcutaneous injection of TNP-KLH – promotes a Th1 response 7 days later (MCKNIGHT et al. 1994). These results support the conclusion that IL-12 alone, with an antigen, is sufficient to promote a stable Th1 response. However, subsequent experiments have clouded this issue somewhat. For example, when mice given a primary immunization with TNP-KLH and IL-12 were rested for 30 days and then rechallenged with TNP-KLH somewhat different results were obtained (BLISS et al. 1996b). Although these animals still exhibited an IFN-γ response, spleen cells from these mice produced IL-4 levels greater than that observed when mice were immunized with TNP-KLH alone. This observation has been interpreted as evidence that IL-12 can act as an effective adjuvant for both Th1 and Th2 cells, although without protection studies – which are not possible in this model – it is unclear how biologically significant the levels of IL-4 observed are in this system. This leaves unresolved the issue of whether IL-12 acts as an effective adjuvant for both Th1 and Th2 type responses in vivo. Certainly all of the in vitro evidence, and most of the in vivo evidence, suggests that IL-12 primarily augments Th1 responses. Other studies with defined antigens have shown that IL-12 does enhance the humoral immune response, although the isotypes involved are those associated with a Th1 type response. Thus, mice immunized with KLH-phospholipase A2 adsorbed to alum and IL-12 exhibited increased levels of IgG2a, IgG2b and IgG3 antibodies, but no increase in IgG1 and a decrease in IgE responses (GERMANN et al. 1995). Nevertheless, the TNP-KLH experiments suggest that under certain circumstances IL-12 will promote a Th2 response. One interpretation of the TNP-KLH results is that they suggest that IL-12 and a single antigen are required, but not sufficient, to generate a stable and strong Th1 response. In all of these systems it is difficult to determine what levels of IFN-γ constitute a strong Th1 response. Alternatively, the ability of IL-12 to act alone as an adjuvant solely for Th1 responses may depend on the natural bias of the immunogen, i.e., whether it favors a Th1 or a Th2 response, or neither, when given without IL-12. Thus, when used as an adjuvant with an immunogen biased towards a Th2 response, which is the case with TNP-KLH, additional cofactors may be required to "lock in" a dominant Th1 response, while in the case of immunogens that have no bias or are slightly biased towards a Th1 response, such cofactors may be less important.

The recent understanding of the role of IL-12 in modulating the response to orally administered antigens may provide new approaches for the use of IL-12 as an adjuvant in vaccination. Feeding high doses of antigens such as ovalbumin induces tolerance in the peripheral lymphoid tissue marked by suppressed proliferative and cytokine response: systemic anti-IL-12 treatment was associated with increased production of TGF-β and T cell apoptosis in both Peyer's patches and peripheral lymphoid tissues, suggesting that IL-12 negatively regulates two of the main mechanisms of oral tolerance (MARTH et al. 1996). Systemic treatment with IL-12 of mice immunized orally with tetanus toxoid stifled the response toward the Th1 type, with increased production of IFN-γ and IL-2, decreased production of Th2 cytokines, increased delayed-type hypersensitivity and a shift in serum IgG1 to IgG2a and IgG3 response (MARINARO et al. 1997). Interestingly, oral administration of IL-12 complexed to liposomes resulted in similar, although not identical, effects to the systemic administration, suggesting that IL-12, or in some cases anti-IL-12, may be used to modulate the immune response to oral vaccination (MARINARO et al. 1997).

Several studies suggest that other cytokines may be important in promoting the efficacy of IL-12 as a vaccine adjuvant. The most well studied of these include cytokines such as IL-1, IL-2 and TNF. Each of these can augment IL-12 activity in vitro, and thus have the potential to augment IL-12 activity in vivo. In the leishmanial system, it was found that depletion of IL-2 compromised Th1 development and inhibited NK cell activation (SCHARTON and SCOTT 1993). Other factors that may contribute to the efficacy of IL-12 as an adjuvant are the costimulatory molecules, B7.1 and B7.2. The ability of B7 to synergize with IL-12 has been shown in vitro, both with human and murine cells (MURPHY et al. 1994; KUBIN et al. 1994b). Moreover, the efficacy of incorporating B7 into a vaccine has been amply demonstrated in murine tumor systems, discussed below (COUGHLIN et al. 1995; ZITVOGEL et al. 1996; RAO et al. 1996).

9 Use of Interleukin-12 in Tumor Therapy

Cytokine based strategies aimed at inducing antitumor immunological responses are being explored due to their potential for systemic effectiveness, modest toxicity and durability of response. Among the cytokines tested, one of the most promising is IL-12 which has proven activity against a variety of murine tumors either as the recombinant protein or as a cytokine secreted by genetically engineered tumor cells. Use of recombinant IL-12 is particularly appealing for clinical translation as it avoids the need for gene transfer. IL-12 by itself is effective against many tumors (BRUNDA et al. 1993; NASTALA et al. 1994; TAHARA et al. 1995; RODOLFO et al. 1996). The antitumor effect of IL-12, when used as a single agent, is mostly mediated by its ability to induce high levels of IFN-γ, which has a direct antitumor effect (NASTALA et al. 1994; ZOU et al. 1995), or activate cells with antitumor activity, such as macrophages, which induce nitric oxide production (WIGGINTON et al. 1996b) and the production of factor with anti-angiogenesis activity (VOEST et al. 1995) such as interferon inducible protein 10 (IP10) (SGADARI et al. 1996). These antitumor effects of IL-12 are usually observed at high doses (e.g. 0.1–1μg daily), close to the maximum tolerated dose for the mice, and with not insignificant toxic effects (BRUNDA et al. 1993). The cells responsible for IFN-γ production and anti-tumor effects in response to IL-12 have been shown in most experimental systems to be primarily CD8[+] T cells, with the participation of NK cells and CD4[+] T cells (BRUNDA et al. 1993; NASTALA et al. 1994; COUGHLIN et al. 1995). In order to optimize the antitumor effect of IL-12 without increasing the dosage to unacceptable toxic levels, various strategies have been used, mostly based on local delivery in order to obtain the highest IL-12 concentration at tumor sites or by combining IL-12 treatment with other factors able to synergize with IL-12 in its antitumor action. A high local concentration of IL-12 has been obtained by peritumoral injection (BRUNDA et al. 1993) or by various gene therapy approaches, including injection into the tumor of IL-12 transfected fibroblasts (ZITVOGEL et al. 1995), use of IL-12 transfected tumor cells (MARTINOTTI et al. 1995; TAHARA et al. 1995; COLOMBO et al. 1996), and local delivery of the IL-12 genes as naked DNA (TAN et al. 1996; RAKHMILEVICH et al. 1996) or in various retroviral, adenovirus, or vaccinia vectors (BRAMSON et al. 1996; CARUSO et al. 1996; MEKO et al. 1996). Among the various combination therapies, the combination of IL-12 with B7-transfected tumor cells (COUGHLIN et al. 1995; ZITVOGEL et al. 1996; RAO et al. 1996), or its association with IL-2 (WIGGINTON et al. 1996a; VAGLIANI et al. 1996), have been proven particularly successful. Indeed, these in vivo successful treatments reflect the well known ability of IL-12 to synergize with IL-2 (CHAN et al. 1992) or with costimulation by B7 (KUBIN et al. 1994b) to induce T cells to produce IFN-γ and other cytokines. Furthermore, IL-12 and B7 costimulation have been shown in vitro to synergize to induce T cell proliferation and efficient generation of cytotoxic CD8[+] T cells (KUBIN et al. 1994b; GAJEWSKI et al. 1995).

Although, in most experimental systems in which IL-12 induced tumor regression, a tumor specific immunity was observed conferring resistance to a second challenge with the same tumors, it is not clear whether the high repeated doses of IL-12 used for obtaining an efficacious nonspecific antitumor effect are the ideal ones in inducing and enhancing the antigen-specific memory response against the tumors. The hematological toxicity of high doses of IL-12, which include a lymphoid toxicity, may alter the immunopotentiating ability of IL-12, and we (KURZAWA et al. 1998a,b) have indeed observed that treatment with high doses of IL-12 prevents the generation of tumor-specific cytotoxic T lymphocytes induced by immunization with granulocyte/macrophage colony-stimulating factor (GM-CSF)-transfected tumor cells. In the setting of using a mutated p53 peptide as a therapeutic vaccine against the Meth A sarcoma, IL-12, at very low doses but not at high doses, has been shown to have a potent adjuvant activity potentiating generation of CTLs and inducing tumor rejection (NOGUCHI et al. 1995). In another model, IL-12 has been shown to dramatically potentiate the ability of dendritic cells pulsed with a class I-restricted tumor peptide to induce tumor-specific immunity against the P815 tumor (BIANCHI et al. 1996). Thus, it is clear that IL-12 has the potential to be used as an adjuvant in anti-tumor vaccination; however, in a clinical setting, it will be important to establish an appropriate treatment schedule that will combine the immediate anti-tumor effects of IL-12, mostly depending on its pro-inflammatory properties, best observed with long-term treatments at high doses, with its ability to potentiate or induce a specific anti-tumor resistance, depending on its immunomodulatory properties, which are optimal at lower doses and with short-term treatments.

10 Use of Interleukin-12 as an Immunotherapeutic Agent

The biological properties of IL-12 make it an excellent candidate for use as an immunopotentiator, particularly when cell-mediated immune responses are required. Many studies have been done with a variety of infectious diseases and tumors demonstrating that IL-12, when given at the initiation of the infection or inoculation of tumor cells, can promote resistance. However, in clinical situations IL-12 will be administered after the disease is established. In some clinical situations successful use of IL-12 will require augmenting a weak immune response, while in others it may require down-regulating an established Th2 response and up-regulating a Th1 response. While the former may be relatively straightforward, the latter may require a better understanding of the factors that are involved in maintaining a Th2 response.

While many experimental manipulations are capable of directing the immune response in one way or the other at the time of infection with *L. major*, it has been more difficult to alter an established Th2 response in mice. IL-12 can promote Th1 development in the normally susceptible BALB/c mouse when given at the time of parasites inoculation, but not when administered 2 weeks after infection (HEINZEL et al. 1995; SYPEK et al. 1993; NABORS et al. 1995). However, when BALB/c mice are simultaneously treated with IL-12 and a leishmanicidal drug – which decreased the parasite burden but does not eliminate all of the parasites – BALB/c mice were able to heal (NABORS et al. 1995). Moreover, when the immune response was examined these animals had switched from a dominant Th2 to a Th1 response (NABORS et al. 1995). These results strongly suggest that high parasite numbers, or antigen load, can influence whether IL-12 can be effective as an immunotherapy.

IL-12 has also been shown to switch the dominant Th2 response associated with egg deposition in mice infected with the helminth *S. mansoni*. The principal pathology caused by this infection relates to the presence of eggs in the liver by the worms, the subsequent immunologic response leading to the development of a granuloma around each egg, and the resulting fibrosis. These granulomas are characterized by the presence of high numbers of eosinophils and are considered a Th2 type response. IL-12 treatment of animals prior to egg injection dramatically reduced granuloma size, and could even control granuloma development if the animals had been previously sensitized to eggs (WYNN et al. 1994). The control of granuloma size extended to a reduction in tissue fibrosis, demonstrating that IL-12 can act as an adjuvant for a vaccine that might control pathology (WYNN et al. 1995a).

Another model in which IL-12 replacement therapy has been able to affect an established infection has been genetically in susceptible BALB/c mice infected with *Mycobacterium avium* (KOBAYASHI et al. 1996). In this model, mice were treated with daily doses of IL-12 for 3 weeks starting 21 days after inoculation with 10^8 cfu of *M. avium*; IL-12 inhibited in vivo mycobacterial growth in a dose-dependent manner, and the growth of mycobacteria in the infected mice remained persistently reduced up to 100 days after IL-12 treatment (KOBAYASHI et al. 1996).

Although in most clinical situations, IL-12 treatment will be useful only if effective against an established infection, prophylactic treatment may be advantageous when it is known that the individuals will be exposed to the pathogens or when specific pathological conditions will result in a depressed production of IL-12 and increased susceptibility to bacterial challenge. The prophylactic effect of IL-12 treatment alone in preventing malaria infection, demonstrated both in mice (SEDEGAH et al. 1994) and in monkeys (HOFFMAN et al. 1997), represents an example of a possible clinical use of IL-12. An example of a pathological condition, associated with increased susceptibility to infection and in which IL-12 treatment has been shown to increase survival, is after a septic challenge in severe burn or traumatic injury (O'SUILLEABHAIN et al. 1996; O'SULLIVAN et al. 1995). This suggests that IL-12 replacement therapy may have a similar prophylactic effect against opportunistic infection in those virus infections which provoke a deficient IL-12 production and either a chronic (e.g., HIV; CHEHIMI et al. 1994) or transient (e.g., measles virus; KARP et al. 1996) immunodepression.

A possible use of the ability of IL-12 to induce a shift from Th2 to Th1 response, in addition to inducing protective Th1 response to pathogens, could be that of preventing or suppressing unwanted Th2 responses, e.g., those observed in allergic response. Although IL-12 has been shown in vitro to shift the human clonal T cell response to allergen from Th2 to Th1 (MANETTI et al. 1993), the analysis of the efficacy of IL-12 in the treatment of allergic disorders has been mostly limited to the study of atopic asthma in mice, in which IL-12 was shown to inhibit antigen-induced airway hyperresponsiveness, inflammation, and Th2 cytokine expression (GAVETT et al. 1995; KIPS et al. 1996). Interestingly, this suppressive effect of IL-12 was observed even with already established Th2 responses, indicating it may provide a novel immunotherapy for the treatment of pulmonary allergic disorders.

11 Interleukin-12: Future Use in Clinical Settings

The preclinical results with IL-12 strongly suggest that this cytokine has a strong potential to be utilized as a therapeutic agent in many clinical conditions, including infectious diseases, severe allergy, and tumors. The ability of IL-12 to act as a powerful adjuvant in immunization, able, in some experimental conditions, even to revert existing states of tolerance or anergy, makes it a powerful candidate to be used in the composition of vaccines for infectious agents or tumors, both pro-phylactic and, most important in the case of tumors, therapeutic. Phase I clinical trials have been started in the last few years in oncology (with an emphasis on melanomas, renal cell carcinomas and cutaneous cell lymphomas), HIV infection, Kaposi's sarcoma, and chronic hepatitis B and C. Although most investigators agree that the use of IL-12 as an adjuvant is the one that offers the greatest potential, clinical attempts have been very limited and only in the oncology setting. Because of the obvious difficulties in treating a rarely life-threatening situation such as allergy with a potent drug such as IL-12, the use of IL-12 in the therapy of severe allergy or asthma still remains a theoretical possibility.

The results of these clinical trials are largely unpublished and only fragmentary information has been presented at a few meetings. Besides a few anecdotal reports, it is too early, especially in the phase I trials, to determine efficacy. In most trials the maximum tolerated doses were determined and, although considerable toxicity was observed at doses lower than expected from the results in experimental animals, schedule and doses for phase II studies were established. Similarly to what was observed in experimental animals (CAR et al. 1995), in high dose treatments acute toxicity was observed in the hematopoietic system, in the liver and in the gastro-intestinal tract, and it was likely dependent on IFN-γ. In long-term treatments toxicity was mostly pulmonary and, by analogy with the experimental animal ev-idence, probably independent of IFN-γ (CAR et al. 1995).

In one of the earliest oncology phase II trials, using daily i.v. doses of IL-12, severe toxicity requiring hospitalization and leading to the death of two patients was observed in the majority of the patients initially entered in the protocol. This toxicity, unexpected on the basis of the phase I results, was attributed to a change in the schedule between phase I and phase II, specifically to the fact that a single dose, 2 weeks before the initiation of the daily i.v. treatment, was part of the phase I protocol and omitted in the phase II. Indeed, subsequent experimentation has shown that mice can be efficiently protected from lethal IL-12 doses, administered by daily treatment, if a single high dose is delivered 1 week before initiation of the daily injections (LEONARD et al. 1997; COUGHLIN et al. 1997). These unexpected difficulties reflect the complex biology of IL-12 and its dual role in vivo as a proinflammatory cytokine as well as an immunomodulatory factor. These tempo-rary setbacks reflect the efficacy of these molecules and, rather than discourage the attempts to use it clinically, should push us to better understand its in vivo effects in order to establish the best schedule and modality of treatment that could harness its full therapeutic potential.

Acknowledgements. We thank Marion Sacks for typing the manuscript. The experimental work described in this chapter was supported by NIH grants CA10815, CA20833, CA32898, AI34412 and AI35914.

References

Afonso LCC, Scharton TM, Vieira LQ, Wysocka M, Trinchieri G, Scott P (1994) The adjuvant effect of interleukin-12 in a vaccine against *Leishmania major*. Science 263:235 237

Aste-Amezaga M, D'Andrea A, Kubin M, Trinchieri G (1994) Cooperation of natural killer cell stimulatory factor/interleukin-12 with other stimuli in the induction of cytokines and cytotoxic cell-associated molecules in human T and NK cells. Cell Immunol 156:480 492

Bacon CM, McVicar DW, Ortaldo JR, Rees RC, O'Shea JJ, Johnston JA (1995a) Interleukin 12 (IL-12) induces tyrosine phosphorylation of JAK2 and TYK2: differential use of janus family tyrosine kinases by IL-2 and IL-12. J Exp Med 181:399 404

Bacon CM, Petricoin EF, III, Ortaldo JR, Rees RC, Larner AC, Johnston JA, O'Shea JJ (1995b) IL-12 induces tyrosine phosphorylation and activation of STAT4 in human lymphocytes. Proc Natl Acad Sci USA 92:7307 7311

Benjamin D, Sharma V, Kubin M, Klein JL, Sartori A, Holliday J, Trinchieri G (1996) IL-12 expression in AIDS-related lymphoma B cell lines. J Immunol 156:1626 1637

Bianchi R, Grohmann U, Belladonna ML, Silla S, Fallarino F, Ayroldi E, Fioretti MC, Puccetti P (1996) IL-12 is both required and sufficient for initiating T cell reactivity to a class I-restricted tumor peptide (P815AB) following transfer of P815AB-pulsed dendritic cells. J Immunol 157:1589 1597

Biron CA, Gazzinelli RT (1995) Effects of IL-12 on immune responses to microbial infections: a key mediator in regulating disease outcome. Curr Opin Immunol 7:485 496

Bliss J, Maylor R, Stokes K, Murray KS, Ketchum MA, Wolf SF (1996a) Interleukin-12 as vaccine adjuvant: Characteristics of primary, recall, and long-term resistance. Ann New York Acad Sci 795:26 35

Bliss J, Van Cleave V, Murray K, Wiencis A, Ketchum M, Maylor R, Haire T, Resmini C, Abbas AK, Wolf SF (1996b) IL-12 as an adjuvant, promotes a T helper 1 cell, but does not suppress a T helper 2 cell recall response. J Immunol 156:887 894

Bramson JL, Hitt M, Addison CL, Muller WJ, Gauldie J, Graham FL (1996) Direct intratumoral injection of an adenovirus expressing interleukin-12 induces regression and long-lasting immunity that is associated with highly localized expression of inteleukin-12. Human Gene Ther 7:1995 2002

Brunda MJ, Luistro L, Warrier RR, Wright RB, Hubbard BR, Murphy M, Wolf SF, Gately MK (1993) Antitumor and antimetastatic activity of interleukin-12 against murine tumors. J Exp Med 178:1223 1230

Car BD, Eng VM, Schnyder B, LeHir M, Shakhov AN, Woerly G, Huang S, Aguet M, Anderson TD, Ryffel B (1995) Role of interferon-g in IL-12 induced pathology in mice. Am J Pathol 147:1693 1707

Caruso M, Pham-Nguyen K, Kwong YL, Xu B, Kosai KI, Finegold M, Woo SL, Chen SH (1996) Adenovirus-mediated interleukin-12 gene therapy for metastatic colon carcinoma. Proc Natl Acad Sci USA 93:11302 11306

Cassatella MA, Meda L, Gasperini S, D'Andrea A, Ma X, Trinchieri G (1995) Interleukin-12 production by human polymorphonuclear leukocytes. Eur J Immunol 25:1 5

Cella M, Scheidegger D, Plamer-Lehmann K, Lane P, Lanzavecchia A, Alber G (1996) Ligation of CD40 on dendritic cells triggers production of high levels of interleukin-12 and enhances T cell stimulatory capacity: T-T help via APC activation. J Exp Med 184:747 752

Chan SH, Kobayashi M, Santoli D, Perussia B, Trinchieri G (1992) Mechanisms of IFN-g induction by natural killer cell stimulatory factor (NKSF/IL-12): Role of transcription and mRNA stability in the synergistic interaction between NKSF and IL-2. J Immunol 148:92 98

Chan SH, Perussia B, Gupta JW, Kobayashi M, Pospisil M, Young HA, Wolf SF, Young D, Clark SC, Trinchieri G (1991) Induction of IFN-g production by NK cell stimulatory factor (NKSF): characterization of the responder cells and synergy with other inducers. J Exp Med 173:869 879

Chehimi J, Starr S, Frank I, D'Andrea A, Ma X, MacGregor RR, Sennelier J, Trinchieri G (1994) Impaired interleukin-12 production in human immunodeficiency virus-infected patients. J Exp Med 179:1361 1366

Chua AO, Chizzonite R, Desai BB, Truitt TP, Nunes P, Minetti LJ, Warrier RR, Presky DH, Levine SF, Gately MK, Gubler U (1994) Expression cloning of a human IL-12 receptor component. A new member of the cytokine receptor superfamily with strong homology to gp130. J Immunol 153:128 136

Chua AO, Wilkinson VL, Presky DH, Gubler U (1995) Cloning and characterization of a mouse IL-12 receptor-b component. J Immunol 155:4286 4294

Colombo MP, Vagliani M, Spreafico F, Parenza M, Chiodoni C, Melani C, Stoppacciaro A (1996) Amount of interleukin-12 available at the tumor site is critical for tumor regression. Cancer Res 56:2531 2534

Coughlin CM, Wysocka M, Kurzawa HL, Lee WMF, Trinchieri G, Eck SL (1995) B7-1 and IL-12 synergistically induce effective antitumor immunity. Cancer Res 55:4980 4987

Coughlin CM, Wysocka M, Trinchieri G, Lee WMF (1997) The effect of IL-12 desensitization on the anti-tumor efficacy of recombinant IL-12. Cancer Res 57:2460 2467

D'Andrea A, Aste-Amezaga M, Valiante NM, Ma X, Kubin M, Trinchieri G (1993) Interleukin-10 inhibits human lymphocyte IFN-g production by suppressing natural killer cell stimulatory factor/interleukin-12 synthesis in accessory cells. J Exp Med 178:1041 1048

D'Andrea A, Ma X, Aste-Amezaga M, Paganin C, Trinchieri G (1995) Stimulatory and inhibitory effects of IL-4 and IL-13 on production of cytokines by human peripheral blood mononuclear cells: priming for IL-12 and TNF-a production. J Exp Med 181:537 546

D'Andrea A, Rengaraju M, Valiante NM, Chehimi J, Kubin M, Aste-Amezaga M, Chan SH, Kobayashi M, Young D, Nickbarg E, Chizzonite R, Wolf SF, Trinchieri G (1992) Production of natural killer cell stimulatory factor (NKSF/IL-12) by peripheral blood mononuclear cells. J Exp Med 176:1387 1398

Desai BB, Quinn PM, Wolitzky AG, Mongini PKA, Chizzonite R, Gately MK (1992) The IL-12 receptor. II. Distribution and regulation of receptor expression. J Immunol 148:3125 3132

Eng VM, Car BD, Schnyder B, Lorenz M, Lugli S, Aguet M, Anderson TD, Ryffel B, Quesniaux VFJ (1995) The stimulatory effects of interleukin (IL)-12 on hematopoiesis are antagonized by IL-12 induced interferon gamma in vivo. J Exp Med 181:1893 1898

Flesch IEA, Hess JH, Huang S, Aguet M, Rothe J, Bluethmann H, Kaufmann SHE (1995) Early interleukin 12 production by macrophages in response to mycobacterial infection depends on interferon-g and tumor necrosis factor-a. J Exp Med 181:1615 1621

Gajewski TF, Renauld J, Van Pel A, Boon T (1995) Costimulation with B7-1, IL-6, and IL-12 is sufficient for primary generation of murine antitumor cytolytic T lymphocytes in vitro. J Immunol 154:5637 5648

Gately MK, Desai BB, Wolitzky AG, Quinn PM, Dwyer CM, Podlaski FJ, Familletti PC, Sinigaglia F, Chizzonite R, Gubler U, Stern AS (1991) Regulation of human lymphocyte proliferation by a heterodimeric cytokine, IL-12 (cytotoxic lymphocyte maturation factor). J Immunol 147:874 882

Gately MK, Wolitzky AG, Quinn PM, Chizzonite R (1992) Regulation of human cytolytic lymphocyte responses by interleukin-12. Cell Immunol 143:127 142

Gavett SH, O'Hearn DJ, Li X, Huang SK, Finkelman FD, Wills-Karp M (1995) Interleukin 12 inhibits antigen-induced airway hyperresponsiveness, inflammation, and Th2 cytokine expression in mice. J Exp Med 182:1527 1536

Gazzinelli RT, Wysocka M, Hayashi S, Denkers EY, Hieny S, Caspar P, Trinchieri G, Sher A (1994) Parasite induced IL-12 stimulates early IFN-g synthesis and resistance during acute infection with Toxoplasma gondii. J Immunol 153:2533 2543

Gearing DP, Cosman D (1991) Homology of the p40 subunit of natural killer cell stimulatory factor (NKSF) with the extracellular domain of the interleukin-6 receptor. Cell 66:9 10

Germann T, Bongartz M, Dlugonska H, Hess H, Schmitt E, Kolbe L, Kolsch E, Podlaski FJ, Gately MK, Rude E (1995) Interleukin-12 profoundly up-regulates the synthesis of antigen-specific complement-fixing IgG2a, IgG2b and IgG3 antibody subclasses in vivo. Eur J Immunol 25:823 829

Germann T, Gately MK, Schoenhaut DS, Lohoff M, Mattner F, Fischer S, Jin S, Schmitt E, Rüde E (1993) Interleukin-12/T cell stimulating factor, a cytokine with multiple effects on T helper type 1 (T$_h$1) but not on T$_h$2 cells. Eur J Immunol 23:1762 1770

Gillessen S, Carvajal D, Ling P, Podlaski FJ, Stremlo DL, Familletti PC, Gubler U, Presky DH, Stern AS, Gately MK (1995) Mouse interleukin-12 (IL-12) p40 homodimer: a potent IL-12 antagonist. Eur J Immunol 25:200 206

Gollob JA, Kawasaki H, Ritz J (1997) Interferon-g and interleukin-4 regulate T cell interleukin-12 responsiveness through the differential modulation of high-affinity interleukin-652 receptor expression. Eur J Immunol 27:647 652

Heinzel FP, Rerko RM, Ahmed F, Pearlman E (1995) Endogenous IL-12 is required for control of Th2 cytokine responses capable of exacerbating leishmaniasis in normally resistant mice. J Immunol 155:730 739

Heinzel FP, Rerko RM, Ling P, Hakimi J, Schoenhaut DS (1994) Interleukin 12 is produced in vivo during endotoxemia and stimulates synthesis of gamma interferon. Infect Immun 62:4244 4249

Hoffman SL, Crutcher JM, Puri SK, Ansari AA, Villinger F, Franke ED, Singh PP, Finkelman F, Gately MK, Dutta GP, Sedegah M (1997) Sterile protection of monkeys against malaria after administration of interleukin-12. Nature Med 3:80 83

Hsieh C, Macatonia SE, Tripp CS, Wolf SF, O'Garra A, Murphy KM (1993) *Listeria*-induced Th1 development in ab-TCR transgenic CD4 T cells occurs through macrophage production of IL-12. Science 260:547 549

Jacobsen SEW (1995) IL-12, a direct stimulator and indirect inhibitor of hematopoiesis. Res Immunol 146:506 514

Jacobson NG, Szabo SJ, Weber-Nordt RM, Zhong Z, Schreiber RD, Darnell JE, Jr, Murphy KM (1995) Interleukin 12 signaling in T helper type 1 (Th1) cells involves tyrosine phosphorylation of signal transducer and activator of transcription (Stat)3 and Stat4. J Exp Med 181:1755 1762

Jelinek DF, Braaten JK (1995) Role of IL-12 in human B lymphocyte proliferation and differentiation. J Immunol 154:1606 1613

Julia V, Rassoulzadegan M, Glaichenhaus N (1996) Resistance to *Leishmania major* induced by tolerance to a single antigen. Science 274:421

Karp CL, Wysocka M, Wahl LM, Ahearn JM, Cuomo PJ, Sherry B, Trinchieri G, Griffin DE (1996) Mechanism of suppression of cell-mediated immunity by measles virus. Science 273:228 231

Kips JC, Brusselle GJ, Joos GF, Peleman RA, Tavernier JH, Devos RR, Pauwels RA (1996) Interleukin-12 inhibits antigen-induced airway hyperresponsiveness in mice. Am J Resp Crit Care Med 153:535 539

Kobayashi K, Yamazaki J, Kasama T, Katsura T, Kasahara K, Wolf SF, Shimamura T (1996) Interleukin (IL)-12 deficiency in susceptible mice infected with Mycobacterium avium and amelioration of established infection by IL 12 replacement therapy. J Infect Dis 174:564 573

Kobayashi M, Fitz L, Ryan M, Hewick RM, Clark SC, Chan S, Loudon R, Sherman F, Perussia B, Trinchieri G (1989) Identification and purification of Natural Killer cell stimulatory factor (NKSF), a cytokine with multiple biologic effects on human lymphocytes. J Exp Med 170:827 846

Kubin M, Chow JM, Trinchieri G (1994a) Differential regulation of interleukin-12 (IL-12), tumor necrosis factor-a, and IL-1b production in human myeloid leukemia cell lines and peripheral blood mononuclear cells. Blood 83:1847 1855

Kubin M, Kamoun M, Trinchieri G (1994b) Interleukin-12 synergizes with B7/CD28 interaction in inducing efficient proliferation and cytokine production of human T cells. J Exp Med 180:211 222

Kurzawa H, Wysocka M, Aruga E, Chang AE, Trinchieri G, Lee WMF (1998a) Recombinant interleukin-12 enhances cellular immune responses to vaccinatiion only after a period of suppression. Cancer Res 58:491 499

Kurzawa H, Hunter CA, Wysocka M, Trinchieri G, Lee WMF (1998b) Immune suppression by recombinant IL-12 involves IFNγ induction of in iNOS activity: inhibitors of NO generation reveal the extent of rIL-12 vaccine adjuvant effect. J Exp Med in press

Leonard JP, Sherman ML, Fisher GL, Buchanan LJ, Larsen G, Atkins MB, Sosman JA, Dutcher JP, Vogelzang NJ, Ryan JL (1997) Effects of single-dose interleukin-12 exposure on interleukin-12-associated toxicity and interferon-gamma production. Blood 90:2541 2548

Leung DYM, Gately M, Trumble A, Ferguson-Darnell B, Schlievert PM, Picker LJ (1995) Bacterial superantigens induce T cell expression of the skin-selective homing receptor, the cutaneous lymphocyte-associated antigen, via stimulation of interleukin 12 production. J Exp Med 181:747 753

Ling P, Gately MK, Gubler U, Stern AS, Lin P, Hollfelder K, Su C, Pan YC, Hakimi J (1995) Human IL-p40 homodimer binds to the IL-12 receptor but does not mediate biologic activity. J Immunol 154:116 127

Locksley RM, Scott P (1991) Helper T-cell subsets in mouse leishmaniasis: induction, expansion and effector function. Immunol Today 12:58 61

Ma X, Aste-Amezaga M, Gri G, Gerosa G, Trinchieri G (1997a) Immunomodulatory functions and molecular regulation of interleukin-12. Chem Immunol 68:1 21

Ma X, Chow JM, Gri G, Carra G, Gerosa F, Wolf SF, Dzialo R, Trinchieri G (1996) The interleukin-12 p40 gene promoter is primed by interferon-g in monocytic cells. J Exp Med 183:147 157

Ma X, Neurath M, Gri G, Trinchieri G (1997b) Identification and characterization of a novel ets-2-related nuclear complex implicated in the activation of the human IL-12 p40 gene promoter. J Biol Chem 272:10389 10401

Macatonia SE, Hosken NA, Litton M, Vieira P, Hsieh C, Culpepper JA, Wysocka M, Trinchieri G, Murphy KM, O'Garra A (1995) Dendritic cells produce IL-12 and direct the development of Th1 cells from naive CD4 ⁺ T cells. J Immunol 154:5071 5079

Mahon BP, Ryan MS, Griffin F, Mills KH (1996) Interleukin-12 is produced by macrophages in response to live or killed Bordetella pertussis and enhances the efficacy of an acellular pertussis vaccine by promoting induction of Th1 cells. Infect Immun 64:5295 5301

Manetti R, Gerosa F, Giudizi MG, Biagiotti R, Parronchi P, Piccinni M, Sampognaro S, Maggi E, Romagnani S, Trinchieri G (1994) Interleukin-12 induces stable priming for interferon-g (IFN-g) production during differentiation of human T helper (Th) cells and transient IFN g production in established Th2 cell clones. J Exp Med 179:1273 1283

Manetti R, Parronchi P, Giudizi MG, Piccinni M-P, Maggi E, Trinchieri G, Romagnani S (1993) Natural killer cell stimulatory factor (NKSF/IL-12) induces Th1-type specific immune responses and inhibits the development of IL-4 producing Th cells. J Exp Med 177:1199 1204

Marinaro M, Boyaka PN, Finkelman FD, Kiyono H, Jackson RJ, Jirillo E, McGhee JR (1997) Oral but not parental interleukin (IL)-12 redirects T helper 2 (Th2)-type responses to an oral vaccine without altering mucosal IgA. J Exp Med 185:415 427

Marth T, Strober W, Kelsall BL (1996) High dose oral tolerance in ovalbumin TCR-transgenic mice: systemic neutralization of IL-12 augments TGF-beta secretion and T cell apoptosis. J Immunol 157:2348 2357

Martinotti A, Stoppacciaro A, Vagliani M, Melani C, Spreafico F, Wysocka M, Parmiani G, Trinchieri G, Colombo M (1995) CD4 T cells inhibit in vivo CD8-mediated immune response against a murine colon carcinoma transduced with IL-12 genes. Eur J Immunol 25:137 146

Maruo S, Oh Hora M, Ahn H, Ono S, Wysocka M, Kaneko Y, Yagita H, Okumura K, Kikutani H, Kishimoto T, Kobayashi M, Hamaoka T, Trinchieri G, Fujiwara H (1997) B cells regulate CD40 ligand-induced IL-12 production in antigen presenting cells (APC) during T cell/APC interactions. J Immunol 158:120 126

Mattner F, Fischer S, Guckes S, Jin S, Kaulen H, Schmitt E, Rüde E, Germann T (1993) The interleukin-12 subunit p40 specifically inhibits effects of the interleukin-12 heterodimer. Eur J Immunol 23:2202 2208

McKnight AJ, Zimmer GJ, Fogelman I, Wolf SF, Abbas AK (1994) Effects of IL-12 on helper T cell-dependent immune responses in vivo. J Immunol 152:2172 2179

Meko JB, Tsung K, Norton JA (1996) Cytokine production and antitumor effect of a nonreplicating, noncytopathic recombinant vaccinia virus expressing interleukin-12. Surgery 120:274 282

Miller MA, Skeen MJ, Ziegler HK (1995) Nonviable bacterial antigens administered with IL-12 generate antigen-specific T cell responses and protective immunity against Listeria monocytogenes. J Immunol 155:4817 4828

Morris SC, Madden KB, Adamovicz JJ, Gause WC, Hubbard BR, Gately MK, Finkelman FD (1994) Effects of IL-12 on in vivo cytokine gene expression and Ig isotype selection. J Immunol 152:1047 1056

Mosmann TR, Coffman RL (1989) TH1 and TH2 cells: different patterns of lymphokine secretion lead to different functional properties. Ann Rev Immunol 7:145 173

Mougneau E, Altare F, Wakil AE, Zheng S, Coppola T, Wang Z, Waldmann R, Locksley RM, Glaichenhaus N (1995) Expression cloning of a protective *Leishmania* antigen. Science 268:563 566

Mountford AP, Anderson S, Wilson RA (1996) Induction of Th1 cell-mediated protective immunity to Schistosoma mansoni by co-administration of larval antigens and IL-12 as an adjuvant. J Immunol 156:4739 4745

Murphy EE, Terres G, Macatonia SE, Hsieh C, Mattson J, Lanier L, Wysocka M, Trinchieri G, Murphy K, O'Garra A (1994) B7 and IL-12 cooperate for proliferation and IFN-g production by mouse T helper clones that are unresponsive to B7 costimulation. J Exp Med 180:223 231

Murphy TL, Cleveland MG, Kulesza P, Magram J, Murphy KM (1995) Regulation of interleukin 12 p40 expression through an NF-*k*B half-site. Mol Cell Biol 15:5258 5267

Nabors GS, Afonso LC, Farrell JP, Scott P (1995) Switch from a type 2 to a type 1 T helper cell response and cure of established *Leishmania major* infection in mice is induced by combined therapy with interleukin 12 and Pentostam. Proc Natl Acad Sci USA 92:3142 3146

Nastala CL, Edington HD, McKinney TG, Tahara H, Nalesnik MA, Brunda MJ, Gately MK, Wolf SF, Schreiber RD, Storkus WJ, Lotze MT (1994) Recombinant IL-12 administration induces tumor regression in association with IFN g production. J Immunol 153:1697 1706

Noguchi Y, Richards EC, Chen Y, Old LJ (1995) Influence of IL-12 on p53 peptide vaccination against established Meth A sarcoma. Proc Natl Acad Sci USA 92:2219 2223

O'Suilleabhain C, O'Sullivan ST, Kelly JL, Lederer J, Mannick JA, Rodrick ML (1996) Interleukin-12 treatment restores normal resistance to bacterial challenge after burn injury. Surgery 120:290 296

O'Sullivan ST, Lederer JA, Horgan AF, Chin DH, Mannick JA, Rodrick ML (1995) Major injury leads to predominance of the T helper-lymphocyte phenotype and diminished interleukin-12 production associated with decreased resistance to infection. Ann Surg 222:482 490

O'Toole M, Wooters J, Brown E, Swiniarski H, Cull G, Leger L, Herrmann S (1996) Interleukin-12 as an adjuvant in peptide vaccines. Ann New York Acad Sci 795:379 381

Ozmen L, Pericin M, Hakimi J, Chizzonite RA, Wysocka M, Trinchieri G, Gately M, Garotta G (1994) IL-12, IFN-g and TNF-a are the key cytokines of the generalized Shwartzman reaction. J Exp Med 180:907 916

Perussia B, Chan S, D'Andrea A, Tsuji K, Santoli D, Pospisil M, Young D, Wolf S, Trinchieri G (1992) Natural killer cell stimulatory factor or IL-12 has differential effects on the proliferation of TCRab+,TCRgd+ T lymphocytes and NK cells. J Immunol 149:3495 3502

Pignata C, Sanghera JS, Cossette L, Pelech S, Ritz J (1994) Interleukin-12 induces tyrosine phosphorylation and activation of 44-kD mitogen-activated protein kinase in human T cells. Blood 83:184 190

Presky DH, Yang H, Minetti LJ, Chua AO, Nabavi N, Wu CY, Gately MK, Gubler U (1996) A functional interleukin 12 receptor complex is composed of two beta-type cytokine receptor subunits. Proc Natl Acad Sci USA 93:14002 14007

Rabinowich H, Herberman RB, Whiteside TL (1993) Differential effects of IL-12 and IL-2 on expression and function of cellular adhesion molecules on purified human natural killer cells. Cell Immunol 152:481 498

Rakhmilevich AL, Turner J, Ford MJ, McCabe D, Sun WH, Sondel PM, Grota K, Yang NS (1996) Gene gun-mediated skin transfection with interleukin 12 gene results in regression of established primary and metastatic murine tumors. Proc Nat Acad Sci USA 93:6291 6296

Rao JB, Chamberlain RS, Bronte V, Carroll MW, Irvine KR, Moss B, Rosenberg SA, Restifo NP (1996) IL-12 is an effective adjuvant to recombinant vaccinia virus-based tumor vaccines: enhancement by simultaneous B7-1 expression. J Immunol 156:3357 3365

Reiner SL, Locksley RM (1995) The regulation of immunity to Leishmania major. Ann Rev Immunol 13:151 177

Reiner SL, Wang Z, Hatam F, Scott P, Locksley RM (1993) Common lineage of Th1 and Th2 subsets in leishmaniasis. Science 259:1457 1460

Rodolfo M, Zilocchi C, Melani C, Cappetti B, Arioli I, Parmiani G, Colombo MP (1996) Immunotherapy of experimental metastases by vaccination with interleukin gene-transduced adenocarcinoma cells sharing tumor-associated antigens: comparison between IL-12 and IL-2 gene-transduced tumor cell vaccines. J Immunol 157:5536 5542

Salcedo TW, Azzoni L, Wolf SF, Perussia B (1993) Modulation of perforin and granzyme messenger RNA expression in human natural killer cells. J Immunol 151:2511 2520

Scharton TM, Scott P (1993) Natural killer cells are a source of interferon gamma that drives differentiation of CD4+ T cell subsets and induces early resistance to Leishmania major of mice. J Exp Med 178:567 577

Scharton-Kersten T, Scott P (1995) The role of the innate immune response in Th1 cell development following L. major infection. J Leukocyte Biol 57:515 522

Scharton-Kersten TM, Wynn TA, Denkers EY, Bala S, Grunvald E, Hieny S, Gazzinelli RT, Sher A (1996) In the absence of endogenous IFN-g, mice develop unimpaired IL-12 responses to Toxoplasma gondii while failing to control acute infection. J Immunol 157:4045 4054

Schijns VE, Haagmans BL, Horzinek MC (1995) IL-12 stimulates an antiviral type 1 cytokine response but lacks adjuvant activity in IFN-g-receptor-deficient mice. J Immunol 155:2525 2532

Scott P (1996) T helper cell development and regulation in experimental cutaneous leishmaniasis. Chem Immunol 63:98 114

Sedegah M, Finkelman F, Hoffman SL (1994) Interleukin 12 induction of interferon gamma-dependent protection against malaria. Proc Nat Acad Sci USA 91:10700 10702

Seder RA, Kelsall BL, Jankovic D (1996) Differential roles for IL-12 in the maintenance of immune responses in infectious versus autoimmune disease. J Immunol 157:2745 2748

Sgadari C, Angiolillo AL, Tosato G (1996) Inhibition of angiogenesis by interleukin-12 -is mediated by the interferon-inducible protein 10. Blood 87:3877 3882

Shu U, Kiniwa M, Wu CY, Maliszewski C, Vezzio N, Hakimi J, Gately M, Delespesse G (1995) Activated T cells induce interleukin-12 production by monocytes via CD40 CD40 ligand interaction. Eur J Immunol 25:1125 1128

Stern AS, Podlaski FJ, Hulmes JD, Pan YE, Quinn PM, Wolitzky AG, Familletti PC, Stremlo DL, Truitt T, Chizzonite R, Gately MK (1990) Purification to homogeneity and partial characterization of cytotoxic lymphocyte maturation factor from human B-lymphoblastoid cells. Proc Natl Acad Sci USA 87:6808 6812

Sypek JP, Chung CL, Mayor SEH, Subramanyam JM, Goldman SJ, Sieburth DS, Wolf SF, Schaub RG (1993) Resolution of cutaneous leishmaniasis: interleukin-12 initiates a protective T helper type 1 immune response. J Exp Med 177:1797 1802

Szabo SJ, Dighe AS, Gubler U, Murphy KM (1997) Regulation of the interleukin (IL)-12R b2 subunit expression in developing T helper (Th1) and Th2 cells. J Exp Med 185:817 824

Tahara H, Zitvogel L, Storkus WJ, Zeh HJ, III, McKinney TG, Schreiber RD, Gubler U, Robbins PD, Lotze MT (1995) Effective eradication of established murine tumors with IL-12 gene therapy using a polycistronic retroviral vector. J Immunol 154:6466 6474

Tan J, Newton CA, Djeu JY, Gutsch DE, Chang AE, Yang NS, Klein TW, Hua Y (1996) Injection of complementary DNA encoding interleukin-12 inhibits tumor establishment at a distant site in a murine renal carcinoma model. Cancer Res 56:3399 3403

Tang YW, Graham BS (1995) Interleukin-12 treatment during immunization elicits a T helper cell type 1-like immune response in mice challenged with respiratory syncytial virus and improves vaccine immunogenicity. J Infect Dis 172:734 738

Trinchieri G, Peritt D, Gerosa F (1996) Acute induction and priming for cytokine production in lymphocytes. Cytokine Growth Factor Rev 7:123 132

Vagliani M, Rodolfo M, Cavallo F, Parenza M, Melani C, Parmiani G, Forni G, Colombo MP (1996) Interleukin 12 potentiates the curative effect of a vaccine based on interleukin 2-transduced tumor cells. Cancer Res 56:467 470

Voest EE, Kenyon BM, O'Reilly MS, Truitt G, D'Amato RJ, Folkman J (1995) Inhibition of angiogenesis in vivo by interleukin 12. J Natl Cancer Inst 87:581 586

Wigginton JM, Komschlies KL, Back TC, Franco JL, Brunda MJ, Wiltrout RH (1996a) Administration of interleukin 12 with pulse interleukin 2 and the rapid and complete eradication of murine renal carcinoma. J Natl Cancer Inst 88:38 43

Wigginton JM, Kuhns DB, Back TC, Brunda MJ, Wiltrout RH, Cox GW (1996b) Interleukin 12 primes macrophages for nitric oxide production in vivo and restores depressed nitric oxide production by macrophages from tumor-bearing mice: implications for the antitumor activity of interleukin 12 and/or interleukin 2. Cancer Res 56:1131 1136

Wolf SF, Temple PA, Kobayashi M, Young D, Dicig M, Lowe L, Dzialo R, Fitz L, Ferenz C, Hewick RM, Kelleher K, Herrmann SH, Clark SC, Azzoni L, Chan SH, Trinchieri G, Perussia B (1991) Cloning of cDNA for natural killer cell stimulatory factor, a heterodimeric cytokine with multiple biologic effects on T and natural killer cells. J Immunol 146:3074 3081

Wu C, Warrier RR, Wang X, Presky DH, Gately MK (1997) Regulation of interleukin receptor beta 1 chain expression and interleukin-12 binding by human peripheral blood mononuclear cells. Eur J Immunol 27:147 154

Wynn TA, Cheever AW, Jankovic D, Poindexter RW, Caspar P, Lewis FA, Sher A (1995a) An IL-12-based vaccination method for preventing fibrosis induced by schistosome infection. Nature 376:594 596

Wynn TA, Eltoum I, Oswald IP, Cheever AW, Sher A (1994) Endogenous interleukin 12 (IL-12) regulates granuloma formation induced by eggs of Schistosoma mansoni and exogenous IL-12 both inhibits and prophylactically immunizes against egg pathology. J Exp Med 179:1551 1561

Wynn TA, Jankovic D, Hieny S, Cheever AW, Sher A (1995b) IL-12 enhances vaccine-induced immunity to Schistosoma mansoni in mice and decreases T helper 2 cytokine expression, IgE production, and tissue eosinophilia. J Immunol 154:4701 4709

Wysocka M, Kubin M, Vieira LQ, Ozmen L, Garotta G, Scott P, Trinchieri G (1995) Interleukin-12 is required for interferon-g production and lethality in lipopolysaccharide-induced shock in mice. Eur J Immunol 25:672 675

Zitvogel L, Robbins PD, Storkus WJ, Clarke MR, Maeurer MJ, Campbell RL, Davis CG, Tahara H, Schreiber RD, Lotze MT (1996) Interleukin-12 and B7.1 costimulation cooperate in the induction of effective antitumor immunity and therapy of established tumors. Eur J Immunol 26:1335 1341

Zitvogel L, Tahara H, Robbins PD, Storkus WJ, Clarke MR, Nalesnik MA, Lotze MT (1995) Cancer immunotherapy of established tumors with IL-12. Effective delivery by genetically engineered fibroblasts. J Immunol 155:1393 1403

Zou J, Yamamoto N, Fujii T, Takenaka H, Kobayashi M, Herrmann SH, Wolf SF, Fujiwara H, Hamaoka T (1995) Systemic administration of rIL-12 induces complete tumor regression and protective immunity: response is correlated with a striking reversal of suppressed interferon-g production by anti-tumor T cells. Int Immunol 7:1135 1145

Redirecting Th1 and Th2 Responses in Autoimmune Disease

C.I. Pearson and H.O. McDevitt

1   Introduction   80
2   Multiple Sclerosis and Experimental Autoimmune Encephalomyelitis
     as Th1-Mediated Diseases   80
3   Insulin Dependent Diabetes Mellitus as a Th1-Mediated Disease   81
4   Rheumatoid Arthritis and Collagen-Induced Arthritis as Th1-Mediated Diseases   82
5   Genetic Influences on Th1/Th2 Development in Autoimmunity   83
6   Therapies of Autoimmune Disease   84
7   Exogenous Cytokines and Antibodies to Cytokines as Therapy   84
8   Interleukin-12 Antagonists as Therapy   87
9   Antigen-Specific Therapies of Autoimmunity   89
10   Specific Antigen Therapy reduces Experimental Autoimmune Encephalomyelitis   96
11   Myelin Basic Protein-Specific T Cell Receptor Transgenic Mice   96
12   Apoptosis and Th Differentiation Correlate with Peptide Affinity for MHC   97
13   Altered Peptide Ligands Can Induce Th2 Responses as Therapy
     for Experimental Autoimmune Encephalomyelitis   100
14   Specific Antigen Therapy in Insulin-Dependent Diabetes Mellitus Deviates Response
     from Th1 to Th2   102
15   Oral Tolerance as Specific Antigen Therapy   105
16   Costimulatory Molecules as Targets for Therapy   109
17   Expression of Class II MHC Molecules Protects from Diabetes   110
18   Neonatal Tolerance   111
19   Conclusions   111
References   112

Department of Microbiology and Immunology, Fairchild Building D345, Stanford University Medical Center, Stanford, CA 94305, USA

1 Introduction

The immune system's ability to distinguish between self and non-self relies on the intricate regulation of cellular responses to antigens presented by major histo-compatibility complex (MHC) molecules to T cells. These responses include those of T helper cells, which can be divided into two populations, Th1 and Th2. The cytokines interleukin (IL)-12 and interferon (IFN)-γ induce Th1 differentiation in vitro. Th1 cells secrete primarily IL-2, IFN-γ, and tumor necrosis factor (TNF) and are involved in delayed type hypersensitivity, induction of the antibody isotype IgG2a, and defense against intracellular pathogens, such as viruses. IL-4 is the primary cytokine that induces differentiation of Th2 type responses. Th2 cells se-crete primarily IL-4, IL-5, IL-6, and IL-10, and are involved in allergic reactions, isotype switching to IgG1, IgG2b and IgE, and defense against extracellular pathogens, such as helminths. A number of autoimmune diseases, in which the immune system attacks self tissues causing damage and disease, are strongly as-sociated with the presence of a particular MHC haplotype, such as insulin-dependent diabetes mellitus (IDDM), multiple sclerosis (MS) and rheumatoid ar-thritis (RA) (TISCH and McDEVITT 1996; STEINMAN 1996; FELDMANN et al. 1996). IDDM, MS, and RA, which are mediated by T cells, are thought to occur as a result of highly skewed Th1 responses, whereas protection and recovery from these autoimmune diseases is thought to be a result of prevailing Th2 responses (LIBLAU et al. 1995). In contrast, systemic autoimmune diseases, allergic reactions, and HIV infection are thought to be a result of pathogenic Th2 responses, although in diseases such as system lupus erythematosus and polyarteritis Th2 skewing has not been clearly demonstrated (LIBLAU et al. 1995).

2 Multiple Sclerosis and Experimental Autoimmune Encephalomyelitis as Th1-Mediated Diseases

Much data have emerged recently that indicate that multiple sclerosis is primarily a Th1-mediated disease. Evidence is derived from studies in multiple sclerosis and experimental autoimmune encephalomyelitis (EAE), which is an inflammatory autoimmune disease of the central nervous system (CNS), and resembles MS in both its pathology and symptoms. EAE differs from MS in that animals must be immunized with myelin, components of myelin such as myelin basic protein (MBP) or proteolipid protein (PLP), or peptides derived therefrom, emulsified in complete Freund's adjuvant (CFA). While a number of different cells populate lesions in the CNS, including B cells, macrophages and CD8+ T cells, CD4+ T helper cells probably initiate disease, since purified populations of CD4+ T cell clones can induce EAE (ZAMVIL and STEINMAN 1990). Encephalitogenicity of T cell clones correlates strongly with the production of TNF and lymphotoxin (LT), both Th1-

associated cytokines (POWELL et al. 1990; BARON et al. 1993; KUCHROO et al. 1993; VAN DER VEEN et al. 1993). Increased IL-12 production by PBL from MS patients was found to be dependent on CD4+ T cell CD40 ligand interactions (BALASHOV et al. 1997). Furthermore, protection and recovery from EAE is associated with Th2 responses since transcripts encoding IL-4, IL-10, and transforming growth factor (TGF)-β, which suppresses immune responses, are increased in the CNS during recovery from EAE (KHOURY et al. 1992; KENNEDY et al. 1992). Th2 lines or clones cannot induce EAE (VAN DER VEEN et al. 1993;, BARON et al. 1993), and in fact, can prevent EAE when administered to mice immunized with autoantigen emulsified in CFA (CHEN et al. 1994; KUCHROO et al. 1995; NICHOLSON et al. 1995).

3 Insulin Dependent Diabetes Mellitus as a Th1-Mediated Disease

Similar findings have been observed in IDDM, most of which have been observed in the nonobese diabetic (NOD) mouse, in which diabetes very similar to that in humans arises spontaneously. By 10 weeks of age, both male and female NOD mice display insulitis, or lymphocytic infiltration of the islets of Langerhans. The islets contain the insulin-producing β-cells which are destroyed during a diabetogenic response. In NOD mice, about 80% of females and about 20% of males become hyperglycemic by 30 weeks of age. CD4+ T cell lines or clones have been shown to accelerate disease in nondiabetic female NOD mice, although it is clear that CD8+ T cells also play a role in pathogenesis (HASKINS and WEGMANN 1996). A number of studies indicate that a Th1 response predominates during diabetes. Analysis of T cells from infiltrated islets of prediabetic and diabetic NOD mice reveal a predominantly Th1 pattern of response (SHIMADA et al. 1996; ZIPRIS et al. 1996; PILSTROM et al. 1995). These findings parallel those found in diabetes induced by multiple low doses of streptozotocin. All four cytokines, IL-2, IL-4, TNF-α, and IFN-γ, are expressed in the pancreas; IFN-γ, however, but not IL-4, is limited to intraislet sites (HEROLD et al. 1996). Antibody to IFN-γ prevents disease, and prevention is correlated with enhanced expression of IL-4 (HEROLD et al. 1996). IFN-γ, however, is not necessary for diabetes, as NOD mice which have a disrupted gene for IFN-γ still develop disease, although the time of onset is delayed (HULTGREN et al. 1996). IL-12 expression was also found in islets before and during disease onset (ZIPRIS et al. 1996). Cyclophosphamide treatment of female NOD mice, which accelerates diabetes onset, correlated with an increased expression of IFN-γ and nitric oxide synthase, an enzyme which is up-regulated upon exposure to Th1 cytokines, such as IFN-γ. IFN-γ and TNF-α production was high in humans with IDDM compared to normal controls (KALLMANN et al. 1997).

Many CD4+ or CD8+ T cell lines or clones have been isolated from spleen or islets of NOD mice. While in most cases the antigen for which these T cells are specific is unidentified, some CD4+ T cell clones can induce diabetes in NOD or

NOD-scid/scid mice. Other CD4+ T cells are able to induce diabetes only if transferred with CD8+ T cells from a NOD mouse. CD8+ cells alone are also capable of accelerating diabetes in particular strains of mice (WONG et al. 1996; MORGAN et al. 1996; Blanas et al. 1996). Thus, both CD4+ and CD8+ are thought to be important during disease initiation and progression (HASKINS and WEGMANN 1996). Moreover, of those CD4+ T cells that induce disease, only Th1 cells will induce overt diabetes (PETERSON and HASKINS 1996). In another study, Th1 cells were diabetogenic, while Th2 lines induced peri-insulitis only (HEALEY et al. 1995). Th1 lines derived from transgenic mice expressing a T cell receptor (TCR) specific for an unknown islet antigen accelerated diabetes after adoptive transfer. Th2 lines derived from the same TCR transgenic line were unable to induce diabetes, nor did they afford protection against diabetes when coinjected with the Th1 line (KATZ et al. 1995). These results parallel findings in EAE in which established Th2 lines specific for PLP peptide 139–151 were unable to protect against EAE when coinjected with Th1 lines derived from the same lymph node cell cultures as the Th2 lines (KHORUTS et al. 1995). These results suggest that Th2 cells may not have such a strong role in protection against autoimmune disease; however, Th2 cells may be able to exert suppressive effects on developing, rather than established Th1 responses, or the suppressive or protective Th2 responses may necessarily be of a different antigenic specificity than the deleterious Th1 cells.

4 Rheumatoid Arthritis and Collagen-Induced Arthritis as Th1-Mediated Diseases

In RA, the synovial joints become chronically inflamed and infiltrated by activated T cells, macrophages, and B cells, leading to the destruction of cartilage and bone (FELDMANN et al. 1996). The predominance of a Th1 response in RA has not been so clearly elucidated. A number of studies, however, indicate that T cells from peripheral blood, synovial fluid, and synovial membranes make Th1 type cytokines (AL-JANADI et al. 1996; GERLI et al. 1995; SCHULZE-KOOPS et al. 1995; VAN ROON et al. 1995; QUAYLE et al. 1993; MILTENBERG et al. 1992). The rodent model of RA, collagen-induced arthritis (CIA), which must be induced in susceptible rats or mice by immunizing with type II collagen in CFA, is also considered to be mediated by Th1 type responses. Draining lymph node cells after immunization with collagen in mice express IL-1β and TNF-α at the onset of disease, while neither IL-4 nor IL-10 can be detected. During recovery from disease, however, IL-10 expression is increased, consistent with an initial Th1 response during disease induction and a later Th2 response during recovery (MAURI et al. 1996). Interestingly, IL-10 production by activated T cell populations in synovial membranes from 2 RA patients was high compared to controls (COHEN et al. 1995). The stage of disease, whether active inflammation or remission, was not clear, however. Enhanced IL-12 expression is observed in susceptible H-2^q mice, but not in resistant H-2^b mice, presumably due to differences in background strain expression of IL-12 (SZELIGA et al. 1996). Moreover, cells transduced with a vector encoding IL-13, a cytokine that inhibits Th1 responses, attenuates CIA in mice (BESSIS et al. 1996). Nondepleting antibodies to CD4 induce Th2 responses and protect mice from CIA (CHU and LONDEI 1996), and rats from EAE (STUMBLES and MASON 1995).

5 Genetic Influences on Th1/Th2 Development in Autoimmunity

Evidence has recently accumulated that genetic predisposition influences whether an immune response is directed toward a Th1 or Th2 response. This phenomenon is most aptly illustrated by susceptibility or resistance to a *Leishmania major* infection, which is correlated with a Th1 response in C3H and B10 strains of mice, which resolve the infection, and a Th2 response in Balb/c mice, which eventually succumb (reviewed in SHER and COFFMAN 1992). In a similar fashion, autoimmunity depends in part on genetic predisposition to Th1 or Th2 responses, conferring susceptibility or resistance to disease. In a transgenic model of diabetes, in which mice expressing hemagglutinin from influenza virus in the pancreatic islets are crossed to mice expressing a TCR specific for hemagglutinin, double transgenic mice will develop spontaneous diabetes on the B10 background, but not on the Balb/c background. Analysis of the cytokine production in response to peptide by splenocytes revealed that the ratio of IFN-γ to IL-4 was increased in the B10 background when compared to that of splenocytes from Balb/c (SCOTT et al. 1994). In EAE, susceptibility to MBP-induced disease correlated with the production of Th1 cytokines in response to immunization with MBP in females, but not in males (CUA et al. 1995a). This difference in T helper development between males and females may stem from a defect in antigen-presenting cell (APC) function in young males in the induction of Th1-promoting cytokines (CUA et al. 1995b). In another recent report, T cell lines specific to MBP peptide Ac1–16 and restricted to I-A^k developed into encephalitogenic Th1 or non-encephalitogenic Th2 cells depending on whether the APCs that were used in vitro to generate the lines were from B10.BR or B10.A mice, respectively (CONBOY et al. 1997). An unknown soluble factor, presumably generated by the APCs, controls IFN-γ and TNF-α production by Th1 cells (CONBOY et al. 1997). Genetic susceptibility to Th1 or Th2 responses may also lie in polymorphism of cytokine genes, such as IL-12 (V.K. Kuchroo et al., personal communication).

The ability to respond to IL-12 and thus become a Th1 cell was recently genetically mapped to a locus on mouse chromosome 11 (GORHAM et al. 1996). A cluster of genes important for T cell differentiation, such as IL-3, IL-4, IL-5, and interferon regulatory factor-1, resides at this locus. A locus, *idd4*, that confers susceptibility to IDDM (TODD et al. 1991), and a locus that confers susceptibility to EAE map to the same region (BAKER et al. 1995). In contrast, susceptibility and resistance to *Leishmania major* have been mapped to loci on different chromosomes and to a different region on chromosome 11. Thus, the genetic predisposition toward a Th1 or Th2 response may be controlled by multiple genes which are up- or down-regulated in a complex manner. In addition, genes influencing predisposition to autoimmune disease and parasitic disease may also involve many different loci.

6 Therapies of Autoimmune Disease

As the immune system is composed and regulated by a large variety of cells and molecules, a number of therapies have been successfully devised to block the autoimmune response. These include administration of antibodies specific for MHC molecules, CD4, adhesion molecules and homing receptors. Targeting these molecules, however, has the disadvantage of knocking out the function of a large subset of the immune response, debilitating not only autoreactive T cells, but also those that would respond and protect against invading pathogens. More specific therapies have directed efforts toward the T cell receptor, as the TCR repertoire in homozygous PL/J and B10.PL mice in response to MBP is highly limited to one or two Vα or Vβ gene elements (ACHA ORBEA et al. 1988; URBAN et al. 1988). Prevention or treatment of EAE has been successful using antibodies to specific Vβ segments (ACHA ORBEA et al. 1988; URBAN et al. 1988; ZALLER et al. 1990), and through the immunization of mice with peptides derived from the TCR primarily used by PL/J and B10.PL mice, or vaccination with T cell clones themselves (GAUR et al. 1993; VANDENBARK et al. 1989; OFFNER et al. 1991; HOWELL et al. 1989; LIDER et al. 1988; KUMAR et al. 1995), presumably by eliciting regulatory T cells that are specific for the TCR expressed by the encephalitogenic T cells. The TCR repertoire in MS patients has revealed some restriction in TCR repertoire, but wide variation between studies has been noted (OKSENBERG et al. 1993; KOTZIN et al. 1991; WUCHERPFENNIG et al. 1990). In contrast, the TCR repertoire in response to PLP or MBP in SJL/J mice is highly variable (SAKAI et al. 1988; KUCHROO et al. 1992). While targeting the TCR through these various methods has been successful in EAE, such therapy would be impractical at present for many autoimmune diseases such as diabetes and RA, since no specific autoantigen, and thus no specific TCR, has been identified that is sufficient and necessary for disease induction.

7 Exogenous Cytokines and Antibodies to Cytokines as Therapy

Th2 responses, and in particular, the cytokines IL-4 and IL-10, inhibit inflammatory and Th1 responses. Exogenous IL-4 is required in vitro to induce Th2 cells and exerts a dominant effect over other variables such as antigen dose, and to a certain extent, addition of exogenous Th1 cytokines such as IFN-γ (SEDER and PAUL 1994;

O'GARRA and MURPHY 1996). Thus, exogenous cytokines or antibodies to cytokines as therapy for autoimmune disease has been studied extensively. Direct administration of Th2 cytokines in vivo has been used to prime autoreactive T cells in both EAE and the NOD model of diabetes in order to induce protective Th2 responses. IL-4 prevents EAE (RACKE et al. 1994; SANTAMBROGIO et al. 1995; INOBE et al. 1996) and diabetes when given to prediabetic NOD mice (RAPOPORT et al. 1993). IL-4 was found to be highly protective if given on the day of induction of EAE for 4–6 days, rather than if given at later time points, or from the day of induction for up to 11 or 12 days (RACKE et al. 1994; SANTAMBROGIO et al. 1995). IL-4 administration to mice was found to induce TGF-β secretion and prevent EAE (INOBE et al. 1996). Treatment of cartilage from RA patients with IL-4 reduces damage induced in vitro (VAN ROON et al. 1995).

IL-4 itself has been found to enhance, rather than ameliorate, symptoms in some studies. IL-4 exacerbates experimental autoimmune uveoretinitis in rats if given every other day starting the day before disease induction (RAMANATHAN et al. 1996). IL-4 given exogenously at the time of EAE onset induced by PLP 139–151 or at the time of disease induction markedly exacerbated disease, rather than ameliorating it (O'GARRA et al. 1997). Moreover, antibodies to IL-4 administered at the time of disease onset also worsens disease and had a similar effect when given with soluble peptide, a regimen shown to inhibit disease (O'GARRA et al. 1997). In contrast, neutralizing antibodies to IL-12 were highly effective in preventing disease (O'GARRA et al. 1997; LEONARD et al. 1995). Interestingly, the most effective treatment was a combination of anti-IL-12 antibodies, soluble PLP139–151, and exogenous IL-4, which resulted in almost complete abrogation of EAE (O'GARRA et al. 1997). IL-4 may be important in arthritis as well, as anti-IL-4 antibody reduced the severity (but not the incidence) of collagen-induced arthritis (HESSE et al. 1996). IL-4 has pleiotropic effects, and in fact can enhance Th1 responses by T cell clones (E. Murphy and A. O'Garra, unpublished observations).

Like IL-4, IL-10 and TGF-β can ameliorate or prevent autoimmune disease. IL-10 prevents EAE in Lewis rats when given during the initiation of disease (ROTT et al. 1994), protects from EAE through changes in APC function (CUA et al. 1996), and protects from Staphylococcus enterotoxin B (SEB)- or TNF-induced relapses of EAE in SJL/J mice (CRISI et al. 1995). Recombinant vaccinia viruses that encoded IL-1β, IL-2, or IL-10 given to mice had inhibitory effects on EAE, whereas viruses that encoded IL-4 had no effect (WILLENBORG et al. 1995). IL-13, which suppresses macrophage function, also prevents EAE (CASH et al. 1994). Chronic TGF-β reduces the severity of EAE relapses (RACKE et al. 1993) while antibodies to TGF-β increase the severity (JOHNS and SRIRAM 1993). TGF-β will protect from EAE when given 5–9 days after initiation of disease (SANTAMBROGIO et al. 1995). These studies emphasize the importance of TGF-β in suppression of autoimmunity and correlate with those studies that indicate that protection induced by oral tolerance from autoimmune disease is due to secretion of TGF-β1. IL-10 has also been shown to be effective in preventing diabetes (PENNLINE et al. 1994) when given in daily subcutaneous injections to 9- and 10-week-old NOD mice; IL-10 transduced islet-specific Th1 lymphocytes suppress adoptively transferred diabetes (MORITANI et al. 1996), and combination therapy of IL-4 and IL-10 prevents islet graft rejection in NOD mice (RABINOVITCH et al. 1995, FAUST et al. 1996). IL-10 suppresses CIA and ameliorates sustained disease in rats (PERSSON et al. 1996). IL-10 given on days 0–21 after induction of CIA suppresses disease in DBA/1 mice (WALMSLEY et al. 1996; TANAKA et al. 1996).

In certain circumstances, IL-10 will enhance autoimmune disease. IL-10 expressed under the rat insulin promoter, thus targeting expression to the pancreatic beta cells, will accelerate diabetes when the transgene is expressed on the NOD background or when expressed in mice given low doses of streptozotocin (WOGENSEN et al. 1994; MORITANI et al. 1994; MUELLER et al. 1996). These results have been noted only in circumstances where IL-10 is expressed locally in non-lymphoid cells. The difference between these experiments studying transgenic expression of IL-10 in a local environment and those studying the systemic effects of exogenous IL-10 probably stem from differences in the timing and amount of IL-10 that is present. However, given the fact that both IL-4 and IL-10 can exacerbate disease under certain conditions, a two-fold therapy which not only activates Th2 pathways, but also neutralizes Th1 pathways, may be necessary for the most effective treatment of autoimmune disease using exogenous cytokines and cytokine antagonists.

While Th2 type cytokines have been used successfully if given systemically and at the beginning of disease induction in EAE or CIA, or prior to significant insulitis in the NOD model of diabetes, Th1 cytokines such as IFN-γ and TNF-α have also been used to treat autoimmune disease. Again, timing and route of administration is critical for the effects that are observed. TNF-α, thought to be a strong pro-inflammatory cytokine in autoimmunity, particularly in EAE and RA, has differential effects when given to NOD mice, depending on the time of administration. TNF or lymphotoxin will inhibit diabetes when given to diabetes-susceptible adult mice or rats (JACOB et al. 1990; SATOH et al. 1990; TAKAHASHI et al. 1993; SEINO et al. 1993; YANG et al. 1994). When given for 3 weeks to neonates, TNF will increase the incidence and hasten the onset of diabetes (YANG et al. 1994). Accordingly, anti-TNF antibody will prevent disease if given to neonates (YANG et al. 1994), but will enhance disease if given to adults (COPE et al., unpublished observations). TNF-α transgenic expression in the pancreatic islets elicits insulitis, but does not induce diabetes. Local TNF-α expression may prevent the development of autoreactive islet-specific T cells, as CD8 + diabetogenic cloned T cells from diabetic NOD mice could induce diabetes in the transgenic mice. These results indicate that locally expressed TNF did not inhibit the effect of established effector T cells (GREWAL et al. 1996). In contrast, TNF-α will enhance EAE relapses (CRISI et al. 1995). In addition, soluble TNF receptor or antibodies to TNF administered during EAE induction ameliorate disease (RUDDLE et al. 1990; SANTAMBROGIO et al. 1993a,b; SELMAJ et al. 1995; MARTIN et al. 1995), suggesting that TNF-α may have different roles in the pathogenesis of diabetes and EAE.

As in EAE, TNF appears critical for disease induction, and thus soluble TNF receptor-Ig fusion molecules will suppress CIA, and mice with disrupted TNF receptor genes resist disease (MORI et al. 1996). Anti-TNF antibody treatment will prevent CIA when given at disease onset, but is less effective at ameliorating fully established disease (JOOSTEN et al. 1996). Similarly, anti-TNF antibody treatment in RA patients has yielded promising reductions in symptoms (ELLIOTT et al. 1993; RANKIN et al. 1995).

Currently, IFN-β has the most efficacy of a number of regimens tested in controlling symptoms in MS patients (INTERFERON-β MS STUDY GROUP 1993). Although the mechanism of such therapy is unclear, some studies indicate that IFN-β down-regulates TNF-α production and enhances Th2 responses, such as IL-6 and IL-10 production (REP et al. 1996; BROD et al. 1996). In other studies, administration of IFN-γ in vivo prevented EAE (BILLIAU et al. 1988; VOORTHUIS et al. 1990), and antibodies to IFN-γ enhance severity and incidence of disease in both susceptible and resistant strains of mice (BILLIAU et al. 1988; DUONG et al. 1992; LUBLIN et al. 1993; HEREMANS et al. 1996; KRAKOWSKI and OWENS 1996). In sharp contrast, IFN-γ exacerbated disease in human clinical trials (PANITCH et al. 1987; NORONHA et al. 1992). These paradoxical results may be due to differences in timing of administration, as IFN-γ given to CIA in rats 24h prior to or 4–24 days after induction of disease enhanced disease, whereas if given between 24 and 48h, IFN-γ suppressed disease. Antibody to IFN-γ also had a differential effect depending on the time of administration (JACOB et al. 1989). IFN-γ is important in the pathogenesis of diabetes, as anti-IFN-γ antibodies down regulate the incidence of diabetes, whether in spontaneous model in rats (NICOLETTI et al. 1997) or in cyclophosphamide-induced diabetes (DEBRAY-SACHS et al. 1991; CAMPBELL et al. 1991). Local IFN-γ expression in pancreatic β-cells induces diabetes (SARVETNICK et al. 1990).

A recent study suggests that IFN-γ is necessary for β-cell destruction and development of IDDM in some experimental models. Diabetes is induced in mice transgenic for pancreatic β-cell-specific expression of nucleoprotein or glycoprotein from lymphocytic choriomeningitis virus (LCMV) by infecting mice with LCMV. Transgenic mice that have a disrupted IFN-γ gene do not develop diabetes, possibly because expression of MHC molecules are not up-regulated in this model (VON HERRATH and OLDSTONE 1997). IFN-γ cannot be absolutely necessary, however, for autoimmunity, as targeted disruption of the IFN-γ gene did not affect EAE in severity or incidence (FERBER et al. 1996), nor did it prevent incidence of diabetes in the NOD mouse, although the day of onset was delayed (HULTGREN et al. 1996). Furthermore, a disrupted IFN-γ gene renders resistant Balb/c mice susceptible to EAE (KRAKOWSKI and OWENS 1996), and a disrupted IFN-γ receptor gene exacerbates EAE (WILLENBORG et al. 1996a). The LCMV model of diabetes may be dependent upon IFN-γ for diabetes induction; a spontaneous diabetogenic immune response in NOD mice may depend on other cytokines in addition to IFN-γ to up-regulate MHC molecule expression.

8 Interleukin-12 Antagonists as Therapy

Interleukin-12 induces IFN-γ and TNF-α production and enhances Th1 responses and T cell proliferation (O'GARRA and MURPHY 1996). Primarily made by activated monocytes, macrophages and dendritic cells, IL-12 is a disulfide linked heterodimer composed of a constitutively made p35 subunit and a p40 subunit that is up-regulated upon activation. Addition of IL-12 to in vitro cultures stimulates Th1 production (TREMBLEAU et al. 1995). The importance of IL-12 in autoimmunity is clearly shown in a number of recent studies. If given to NOD female mice, IL-12 increases the incidence of IDDM from 60% to 100% (TREMBLEAU et al. 1995). IL-12 induces more severe CIA (GERMANN et al. 1995) and more severe EAE, which is correlated with increased synthesis of nitric oxide synthase by macrophages (LEONARD et al. 1995, 1996). Mice with a targeted disruption of the IL-12 gene have a reduced incidence and severity of CIA (McINTYRE et al. 1996). Antibodies to IL-12 given to NOD mice prevents IDDM (OGATA et al. 1995), while mice with a targeted disruption of the IL-12 gene on the NOD background do not get insulitis by 12 weeks of age. Cyclophosphamide-induced diabetes incidence in NOD females was reduced to 15% in IL-12 knock-out mice compared to an incidence of 60% in control mice (LAMONT and ADORINI 1996). Given the conflicting differences be-tween IL-12 and IFN-γ treatment in EAE, and that IFN-γ exacerbates MS, IL-12 may be a better target for therapy of autoimmunity.

The timing of administration of IL-12 can affect autoimmune disease differ-ently, as has been shown for administration of IL-4 and IFN-γ. Intermittent, in contrast to continuous, administration of IL-12 to NOD mice decreased the inci-dence of IDDM, however (O'HARA et al. 1996). A number of in vitro studies reveal that IL-12 can induce human T cell lines to make both IFN-γ and IL-10 (PAGANIN et al. 1996); the addition of both IL-4 and IL-12 induces cells that produce IL-4 and IFN-γ (DELESPESSE et al. 1996). IL-12 given in vivo at the time of primary im-munization results in an enhanced Th1 response, but also in Th2 responses (BLISS et al. 1996).

Despite these conflicting results, the potential for developing IL-12 antagonists remains promising. The p40 subunit of IL-12 can form disulfide-linked homodimers $(p40)_2$ that bind to the IL-12 receptor (IL-12R) and antagonize IL-12 function (GILLESSEN et al. 1995). For example, proliferation in response to phorbol ester and IL-12, and enhancement of IFN-γ synthesis and proliferation by IL-12 is inhibited by the p40 homodimer (GERMANN et al. 1995). Chronic relapsing EAE induced by PLP can be inhibited by $(p40)_2$ (MAGRAM et al. 1996). CIA also can be suppressed by $(p40)_2$, but high doses of IL-12 also ameliorate CIA (HESS et al. 1996). Another complication is that $(p40)_2$ given to mice during an immune response enhanced Th1 responses, rather than inducing Th2 responses (PICCOTTI et al. 1997). While p40 can be detected in the serum of both mice and humans (GERMANN et al. 1995; TREMBLEAU et al. 1995), the affinity of human $(p40)_2$ for human IL-12R is only 10%–20% of that of murine $(p40)_2$ for the murine IL-12R, and as such, has about a 10% inhibitory efficacy compared to that of murine $(p40)_2$ (GERMANN et al. 1995). At this point, it is unclear whether the p40 subunit is transducing a signal through the IL-12R; if so, this may explain why results using the $(p40)_2$ differ from those using IL-12 knockout mice or antibodies to IL-12.

In summary, results from treatment of autoimmune disease with exogenous cytokines indicate that exogenous cytokine therapy remains a viable option. Induction of Th2 responses through Th2 type cytokines such as IL-4 and IL-10 in general produce a beneficial effect on EAE, diabetes, and CIA. Although the use of Th1 type cytokines such as TNF-α or IFN-γ has also proven successful, these results are so far limited to animal models of diabetes and EAE, respectively. IFN-γ therapy in MS had the opposite effect as in EAE and increased occurrences of relapses, although another interferon, IFN-β, is now in use as a therapy for MS. Antibodies to Th1 type cytokines, and in particular, anti-TNF, have been successful in EAE, CIA, and RA, although no beneficial effect was seen in two MS patients (VAN OOSTEN et al. 1996). Antibodies to IL-12 have also been shown to be beneficial, as well.

A few studies indicate, however, that caution must be exercised in treating autoimmune disease with cytokines or antibodies to cytokines. First, different effects are observed depending on the reagent, such as antibodies to IL-12 versus the (p40)2 antagonist. Second, although each of these autoimmune diseases is considered to be Th1-mediated, the same cytokine has different effects, depending on the disease. For example, TNF-α given to adult NOD mice prevents diabetes, but will enhance EAE relapses, although this differential effect may be explained by different protocols and the timing of administration. Third, a cytokine such as IL-4 or IL-10 may have opposite effects, depending on when and where it is administered, and whether it is given in conjunction with other therapeutics. Finally, exogenous cytokine or antibody to cytokine therapy that skews the immune response toward a Th2 phenotype may render one susceptible to syndromes that are associated with such a subset, such as allergy. Table 1 summarizes the effects of therapy of various autoimmune diseases in both animals and humans using cytokines and anti-cytokine antibodies.

9 Antigen-Specific Therapies of Autoimmunity

In recent years, blocking a specific population of T cells through the administration of specific antigen has been successful in a number of model diseases, including EAE and diabetes. The advantage of these approaches over exogenous cytokine or antibody therapy is that only those T cells that are specific for a particular autoantigen are targeted. These experiments derive from experiments performed in the past that showed that antigen given in incomplete Freund's adjuvant or in a soluble form (either intravenously or intraperitoneally) produces a state of tolerance such that subsequent immunization with the same antigen in CFA does not induce an immune response (reviewed in WEIGLE 1973; MILLER and MORAHAN 1992). In addition, T cells must receive two signals, one through the peptide/MHC/TCR interaction, and one through costimulatory molecules, primarily B7 on the APC and CD28 on the T cells, in order to become fully activated and functional (reviewed in MUELLER et al. 1989; HARDING et al. 1992). Lack of costimulation inactivates the T cell, putting it into an unresponsive state termed anergy, which can

Table 1. The effect of cytokines on autoimmune disease

Treatment	Disease	Route[a]	Time[b]	Effect[c]	Reference
IL-4	EAE	Systemic	d0–11	Prevents	RACKE et al. 1994
IL-4	EAE	Systemic	d0–4	Prevents	SANTAMBROGIO et al. 1995
			d5–9	No benefit	
IL-4	EAE	Systemic	d0–12	Prevents	INOBE 1996
			d0	Reduces severity	
IL-4	EAE	Vaccinia virus	d0, d6	No benefit	WILLENBORG 1995
IL-4	EAE	Systemic	3–4× over 2 days at d0 or d9–11	Exacerbates	O'GARRA 1997
			+ antigen	Exacerbates	
			+ antigen + anti-IL-12	Prevents	
IL-4	CIA	Systemic	IL-4 added in culture to SFMN and cartilage	Reduces severity of damage	VAN ROON 1995
IL-4	EAU	Systemic	d–1. 0. 2. 4. 6. 8	Exacerbates	RAMANATHAN 1996
IL-4	IDDM	Systemic	2× weekly from 6 to 20 weeks	Prevents	RAPOPORT 1993
Anti-IL-4	CIA	Systemic	d1–2	Reduces severity: mice pristane primed	HESSE 1996
IL-10	EAE	Systemic	d0. 3. 6	Prevents	ROTT 1994
IL-10	EAE	Systemic	d–1. 0. 1 relative to induction of relapse	Prevents	CRISI 1995
IL-10	EAE	Vaccinia virus	d0. d6	Reduces severity	WILLENBORG 1995
IL-10	IDDM	Systemic	2× weekly from 10 to 25 weeks of age	Prevents	PENNLINE 1994
IL-10	IDDM	Transduced Th1 cells	Transfer cells at 8 days of age: induce diabetes with CY at d37	Prevents	MORITANI 1996
IL-10	CIA	Systemic	d12–20	Prevents	PERSSON et al. 1996
IL-10	CIA	Subcutaneous in paws	After onset	Reduces severity	PERSSON et al. 1996
IL-10	CIA	Systemic	After onset	Reduces severity	WALMSLEY et al. 1996
IL-10	CIA	Systemic	d0–21	Prevents	TANAKA et al. 1996
			d0–48	Reduces severity	
			d21–48	No benefit	

IL-10	IDDM	Local pancreatic β cell	Embryonic d10	Induces diabetes	WOGENSEN et al. 1994
IL-10	IDDM	Local pancreatic α cell	Embryonic d10	Induces diabetes	MORITANI et al. 1994
IL-10	IDDM	Local pancreatic β cell	Embryonic d10	Renders non susceptible mice diabetic with streptozotocin	MUELLER et al. 1996
IL-10	IDDM	Local pancreatic β cell	Embryonic d10	LCMV-GP×RIP-IL-10 get higher incidence of diabetes than LCMV-GP mice after LCMV infection	LEE et al. 1994
IL-13	EAE	Vector cells	d0	Prevents	CASH et al. 1994
TGF-β2	EAE	Systemic	d1–5	Prevents	RACKE et al. 1993
TGF-β2	EAE	Systemic	d3–7. 3× week for 5 weeks	Reduces severity	RACKE et al. 1993
TGF-β1	EAE	Systemic	d1–5	No benefit	SANTAMBROGIO 1993, 1995
			d5–9	Prevents	
			d9–13	No benefit	
TGF-β1	CIA	Systemic	d0–4	No benefit	SANTAMBROGIO 1995
			d7–11	No benefit	
			d14–18	Prevents	
Anti-TGF-β1	EAE	Systemic	d2, 4, 6, 8, 10	Exacerbates	JOHNS and SRIRAM 1993
Anti-TGF-β1	EAE	Systemic	d5, 9	Exacerbates	SANTAMBROGIO 1995
IFN-γ	EAE	Systemic	d–1, 0, 1, 3, 5, 7, 9	Prevents	BILLIAU et al. 1988
IFN-γ	EAE	Systemic	Intraventricularly d7 or 8	Prevents	VOORTHUIS et al. 1990
IFN-γ	MS	Systemic	1, 30, or 1000 µg i.v. 2× weekly for 4 weeks	Exacerbates	PANITCH et al. 1987
IFN-γ	CIA	Systemic	d0	Exacerbates	JACOB et al. 1989
			d1–2	Reduces	
			d4–12	Exacerbates	
			d12–24	Exacerbates	
Anti-IFN-γ	EAE	Systemic	d–6, 0, 7, 14	Exacerbates	BILLIAU et al. 1988
Anti-IFN-γ	EAE	Systemic	d0, d7	Exacerbates	DUONG et al. 1992
			d0	Exacerbates adoptive transfer	
			d4	No effect on adoptive transfer	

Table 1. (*cont.*)

Treatment	Disease	Route[a]	Time[b]	Effect[c]	Reference
Anti-IFN-γ	EAE	Systemic	Given during remission	Exacerbates relapses	HEREMANS et al. 1996
Anti-IFN-γ	IDDM	Systemic	2× weekly from d33 to d105	Delays onset	NICOLETTI et al. 1997
Anti-IFN-γ	IDDM	Systemic	d0–1	Reduces incidence	DEBRAY SACHS et al. 1991
Anti-IFN-γ	IDDM	Systemic	d–1, 3, 6, 9, 12	Reduces incidence	CAMPBELL et al. 1991
Anti-IFN-γ	CIA	Systemic	d–2,–1 d4–8 d12–24	Reduces severity Reduces severity Exacerbates	JACOB et al. 1989
IFN-γR– –	EAE	No IFN-γ receptor	NA	Exacerbates	WILLENBORG et al. 1996
IFN-γR– –	EAE	No IFN-γ receptor	NA	No effect	FERBER et al. 1996
IFN-γR– –	IDDM	No IFN-γ	NA	Delays onset	HULTGREN et al. 1996
IFN-γR– –	EAE	No IFN-γ	NA	Converts resistant Balb c to susceptible	KRAKOWSKI and OWENS 1996
Lymphotoxin	IDDM	Systemic	3× week from 4 to 30 weeks old	Reduces incidence in NOD	SEINO et al. 1993
Lymphotoxin	IDDM	Systemic	3× week from 4 to 11 weeks old	Reduces incidence in rats	TAKAHASHI et al. 1993
TNF	IDDM	Systemic	3× week from 9 to 21 weeks old	Reduces incidence	JACOB et al. 1990
TNF	IDDM	Systemic	Every other day from 4 to 7 weeks old	Reduces incidence	YANG et al. 1994
TNF	IDDM	Systemic	2× week, 4–27 weeks old	Reduces incidence in NOD	SATOH et al. 1990
TNF	IDDM	Systemic	Daily from 7 to 16 weeks old	Reduces incidence in rats	RABINOVITCH et al. 1995
TNF	IDDM	Systemic	Every other day from 0 to 21 days old	Increases incidence in NOD	YANG et al. 1994
TNF	EAE	Systemic	During remission	Induces relapses	CRISI et al. 1995

Treatment	Disease	Administration	Protocol	Effect	Reference
Anti-TNF	IDDM	Systemic	2× week from 7 weeks old for 8 weeks	Increased insulitis	JACOB et al. 1992
			Every 2 days from 4–7 weeks old	Reduces incidence	YANG et al. 1994
			Every 2 days from 8 to 11 weeks old	Increases onset and incidence	COPE et al. unpub.
Anti-TNF LT	EAE	Systemic	1 i.p. injection 48 h after transfer	Prevents	RUDDLE et al. 1990
Anti-TNF	EAE	Systemic	d5, d9	Reduces severity	SANTAMBROGIO 1993a, b
Anti-TNF	MS	Systemic	Various	Exacerbates	van OOSTEN et al. 1996
Anti-TNF	CIA	Systemic	1 dose i.p. at d30 or d32	Reduces severity	JOOSTEN et al. 1996
Anti-TNF	RA	Systemic	Two or four doses i.v. One or two doses i.v.	Reduces severity	ELLIOTT et al. 1993 RANKIN et al. 1995
sTNFR-Fc	CIA	Systemic	Every 3 days i.v. from d18 to d36 or daily i.p. from d21 to d28	Prevents	MORI et al. 1996
sTNFR-Fc	EAE	Systemic	Various	Reduces severity	SELMAJ et al. 1995
sTNFR	EAE	Systemic	Every 2–3 days, starting d9	Reduces severity	MARTIN et al. 1995
sTNFR-Fc	IDDM	Transgene	Embryonic d12.5	Protects	HUNGER et al. 1997
IL-12	EAE	Systemic	d0–2 after adoptive transfer	Exacerbates	LEONARD et al. 1995
IL-12	CIA	Systemic	Low or high doses d0–4	Exacerbates	GERMANN et al. 1995
IL-12	IDDM	Systemic	Daily for 5 weeks in 8–10-week-old	Increases incidence, onset	TREMBLEAU et al. 1995
IL-12	IDDM	Systemic	1× weekly for 15 weeks in 9–10-week-old NOD	Reduces incidence	O'HARA et al. 1996
IL-12	CIA	Systemic	High doses d0–14, 21	Reduces severity	HESS et al. 1996
Anti-IL-12	EAE	Systemic	Every 2 days d0–6 Every 2 days d0–12	Delays onset Prevents	LEONARD et al. 1995

Table 1. (*cont.*)

Treatment	Disease	Route[a]	Time[b]	Effect[c]	Reference
Anti-IL-12	EAE	Systemic	1 i.p. dose at disease onset	Reduces severity	O'GARRA et al. 1997
Anti-IL-12	IDDM	Systemic	1 day prior to adoptive transfer	Prevents	OGATA et al. 1995
IL-12---	CIA	No IL-12 expression	NA	Reduces incidence. severity	McINTYRE et al. 1996
IL-12---	IDDM	No IL-12 expression	NA	No insulitis in NOD by 12 weeks; 75% reduction in CY-induced IDDM	LAMONT and ADORINI 1996

EAE, experimental autoimmune encephalomyelitis; IL, interleukin; IFN, interferon; IDDM, insulin-dependent diabetes mellitus; CIA, collagen-induced arthritis; TNF, tumor necrosis factor; sTNFR, soluble TNF receptor; NOD, nonobese diabetic mouse; CY, cyclophosphamide; NA, not available; sTNFR-Fc, soluble TNF receptor-Ig fusion protein.

[c]Prevents: EAE. IDDM or CIA incidence completely or almost completely reduced. Reduces severity: the severity of EAE or CIA is reduced. but the incidence not necessarily affected. Exacerbates: EAE or CIA severity increased. incidence either the same or increased. Reduces incidence: IDDM incidence reduced. Delays onset: IDDM incidence unaffected. but the day of onset of disease is delayed.

[a]Route refers to whether the cytokine or anticytokine was given systemically (systemic) or otherwise noted.

[b]Time refers to when cytokine or anti-cytokine was given. All days are relative to day 0 (d0). the day of immunization in EAE or CIA.

be overcome by the addition of exogenous IL-2. Most of the experiments on anergy have been performed in vitro on established T cell clones, and thus how applicable these studies are to naive T cells in vivo remains unclear.

A number of studies have demonstrated that encephalitogenic autoantigen given in IFA or phosphate-buffered saline (PBS) prevents subsequent induction of EAE (SMILEK et al. 1991; METZLER and WRAITH 1993; GAUR et al. 1992). Other studies have extended these results by treating mice with soluble or tolerogenic forms of whole protein or peptide at the time of disease onset (SMILEK et al. 1991; SAMSON and SMILEK 1995; KARIN et al. 1994; CRITCHFIELD et al. 1994; KUCHROO et al. 1994; FRANCO et al. 1994; NICHOLSON et al. 1995; BROCKE et al. 1996; LIU and WRAITH 1996). Similarly, diabetes has been prevented by intrathymic or intravenous injections of putative autoantigen, GAD65 (TISCH et al. 1993, 1994, 1998) or peptide derived from GAD65 (KAUFMAN et al. 1993; TIAN et al. 1996a 1996b). Insulin has also been used to treat NOD mice either orally (HANCOCK et al. 1995) or using the B chain either subcutaneously or through nasal inhalation (DANIEL and WEGMANN 1996). Heat shock protein (HSP) 60 peptide p277 prevents IDDM in NOD mice (ELIAS and COHEN 1996). Oral insulin, however, does not have an effect on the incidence of diabetes in susceptible rats (MORDES et al. 1996).

The mechanisms behind such therapies have yet to be established, although evidence suggests that T cells exposed to soluble antigen become anergized through the lack of appropriate costimulation (GAUR et al. 1992), or deleted through activation-induced cell death (CRITCHFIELD et al. 1994). Recent experiments have shown that coinjection of whole MBP and antibodies to LFA-1 adhesion molecule or immunization with an HSP65 peptide that is cross-reactive with a myelin protein in IFA prevents EAE; prevention is correlated with a shift in the ratio of IgG2a to IgG1, consistent with immune deviation from Th1 to Th2 (WILLENBORG et al. 1996b; BIRNBAUM et al. 1996). Studies using soluble antigen suggest that soluble peptide or protein induces tolerance in the Th1 compartments, while responses in the Th2 compartment were preserved (BURSTEIN et al. 1992; DE WIT et al. 1992). Theiler's murine encephalomyelitis virus (TMEV) induces a form of EAE very similar to that induced by immunization with MBP or PLP. Recently, intravenous injection of UV-inactivated TMEV or TMEV coupled to splenocytes induces tolerance in the Th1 compartment, and concomitant stimulation of Th2 responses (KARPUS et al. 1994; PETERSON et al. 1993). These studies suggest that Th1 cells are more susceptible to tolerogenic mechanisms than are Th2 cells. In particular, Th2 clones express less Fas ligand than do Th1 clones, rendering Th2 clones resistant to Fas-mediated apoptosis (RAMSDELL et al. 1994). Furthermore, the strength of signal, whether in the form of antigen dose or the degree of costimulatory interaction between T cell and antigen presenting cell, influences Th1 and Th2 development (BRETSCHER et al. 1992; HOSKEN et al. 1995; CONSTANT et al. 1995; PEARSON et al. 1997; RULIFSON et al. 1997). Immunization with soluble antigen may stimulate T cells with higher signals, at least initially following immunization, than does immunization with antigen in CFA, leading to primarily Th2 responses.

10 Specific Antigen Therapy reduces Experimental Autoimmune Encephalomyelitis

Soluble antigen therapy may in fact be more efficient at induction of Th2 responses, as single or multiple doses of peptide or whole protein in several systems preferentially induce Th2 responses. This may stem from the fact that antigen given in CFA rather than in IFA or PBS is delivered in smaller amounts over a longer period of time, thus effectively delivering a Th1-inducing signal. Another possibility that is not mutually exclusive is that the addition of *Mycobacterium* in CFA initiates a specific Th1-response to bacterial antigens that will affect the response to the antigen. CFA induces up-regulation of IL-1 and TNF, which help to induce Th1 responses and activate macrophages to release nitric oxide and cause tissue damage.

In EAE, a stronger antigen stimulus is clearly more effective at ameliorating disease. Several studies have shown that Ac1–11, the N-terminal peptide of myelin basic protein, can effectively prevent induction of or treat disease when given high doses intranasally, intravenously or intraperitoneally (METZLER and WRAITH 1993; SAMSON and SMILEK 1995; LIU and WRAITH 1996). These experiments explored the ability to prevent or treat EAE of Ac1–11 and two analogs of Ac1–11, Ac1–11[4A] and Ac1–11[4Y], in which the native lysine at position four is replaced with alanine or tyrosine, increasing the affinity of the peptide for the class II MHC molecule I-A^u by 50- and 1500-fold over that of Ac1–11, respectively. Both of these peptides stimulate Ac1–11-specific T cells in an enhanced fashion, such that approximately 25 times less Ac1–11[4A] and 1000-fold less Ac1–11[4Y] are needed to achieve maximal T cell activation than Ac1–11.

A single dose of 100 µg of Ac1–11 given intranasally was able to inhibit induction of EAE by 50%, while a single dose of 100 µg of Ac1–11[4A] or Ac1–11[4Y] was able to almost completely inhibit induction of EAE (METZLER and WRAITH 1993). Similarly, treatment of EAE by intravenous injection was effective when six doses of 300 µg of Ac1–11 were given; six doses of 3 µg of Ac1–11[4Y] induced the same degree of efficacy (SAMSON and SMILEK 1995). LIU and WRAITH 1996, showed very similar findings with Ac1–11 and Ac1–11[4Y] given intraperitoneally.

11 Myelin Basic Protein-Specific T Cell Receptor Transgenic Mice

None of these studies show that increasing the peptide affinity for the MHC (and thus the effective antigen dose and presumably the efficacy of treatment) correlated with the induction of protective Th2 type responses. Recently, however, these peptides have been shown to induce different T helper subsets in the periphery when given intravenously to mice expressing a transgenic T cell receptor specific for Ac1–

11 and restricted to I-A" (PEARSON et al. 1997). This TCR comes from a CD4+ encephalitogenic Th1 clone (ZAMVIL and STEINMAN 1990) and is expressed by at least 60% of peripheral T cells in the transgenic mouse model. When TCR transgenic mice are given a single dose of 0.24 to 2.4 mg of Ac1–11, Ac1–11[4A], or Ac1–11[4Y] intravenously, CD4+CD8+ double positive thymocytes undergo apoptosis (LIBLAU et al. 1994; PEARSON et al. 1997). The efficiency of programmed cell death correlates with the affinity of the peptide for the MHC, such that 24 h after peptide injection, Ac1–11 induces approximately a 50% reduction in the number of DP cells, Ac1–11[4A] induces an approximately 50%–75% reduction, and Ac1–11[4Y] about a 75% reduction. Six days after peptide injection Ac1–11 treatment results in a twofold in the total number of thymocytes, Ac1–11[4A] results in a fivefold reduction in the total number, and Ac1–11[4Y] results in up to tenfold reduction in the total number of thymocytes (LIBLAU et al. 1994; PEARSON et al. 1997). Most of the remaining cells are either CD4 single positive (SP), CD8 SP, or double negative thymocytes.

Thymocytes from PBS-injected TCR transgenic mice produce primarily small amounts of IL-2 when stimulated in vitro with Ac1–11, Ac1–11[4A] or Ac1–11[4Y]. Thymocytes from peptide-injected mice produce IFN-γ in amounts inversely correlated with the affinity of the peptide for the MHC. The highest amount of IFN-γ was made by cells from Ac1–11-injected mice, intermediate amounts made by cells from Ac1–11[4A]-injected mice, and the least made by cells from Ac1–11[4Y]-injected mice. Production of IL-4, in contrast, was directly correlated with the affinity of the peptide for the MHC, such that cells from Ac1–11-injected mice made little or no IL-4, cells from Ac1–11[4A]-injected mice made intermediate amounts, and cells from Ac1–11[4Y]-injected mice made the highest amount.

12 Apoptosis and Th Differentiation Correlate with Peptide Affinity for MHC

In addition to the fate of thymocytes following peptide injection into Ac1–11-specific TCR transgenic mice, the fate of peripheral T cells was followed. Peripheral T cells are activated within 4 h following injection of peptide, as measured by up-regulation of CD69 and CD25 and down-regulation of CD62L. This state of activation is maintained based on the affinity of the peptide for the MHC. Proliferative responses in vitro are enhanced relative to those from PBS-injected mice for at least 2 days, and return to what is observed in PBS-injected control mice by day five. No classical anergy is observed throughout the time course. Moreover, although the absolute numbers of CD4+ T cells does not decrease, regardless of the peptide injected, the number of cells undergoing activation-induced cell death 24 h after injection of peptide increases with the affinity of the peptide for the MHC. At the same time, T cells surviving encounter with soluble antigen undergo T helper subset differentiation, paralleling that seen for thymocytes (Fig. 1), an observation

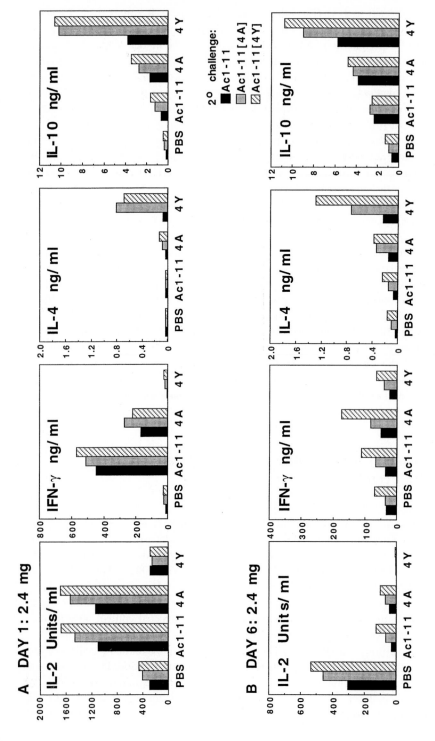

Treatment

that holds for at least 6 days following peptide therapy. Similar results were obtained when TCR transgenic mice were given four daily doses of up to 600 micrograms per dose of Ac1–11, Ac1–11[4A], or as little as 12 micrograms per dose of Ac1–11[4Y]. Thus, in vivo, Ac1–11 is a strong inducer of Th1 responses, while Ac1–11[4Y] is a strong inducer of Th2 responses, with Ac1–11[4A] intermediate in induction of T helper 1 and 2 phenotypes.

These findings were extended to in vitro studies, in which purified CD4+ naive T cells were cultured in vitro with syngeneic irradiated splenocytes and various amounts of Ac1–11, Ac1–11[4A] or Ac1–11[4Y], then restimulated in a secondary challenge with a single dose of peptide. Analysis of the cytokines made in the secondary stimulation revealed that Ac1–11 was able to induce only Th1 type responses, whereas Ac1–11[4A] and Ac1–11[4Y] could induce both Th1 and Th2 type responses, depending on the dose of peptide in the primary stimulation. Ac1–11[4Y] in this case was the most potent at inducing Th2 responses (PEARSON et al. 1997; Fig. 2). These in vitro studies parallel those in vivo and in general other studies which suggest high doses of antigen induce Th2 responses. Interestingly, Ac1–11[4Y] induced Th1 responses when the primary concentration of peptide was very high or very low, and Th2 responses when the primary concentration was intermediate (Fig. 2). These findings differ from other experiments in which naive CD4+ T cells specific for ovalbumin peptide 323–339 (OVA) were stimulated with various amounts of OVA peptide and purified dendritic cells as APC. In these experiments, high and low doses of peptide in the primary stimulation induced Th2 responses, whereas intermediate doses induced Th1 responses (HOSKEN et al. 1995). The apparent discordant findings of these two studies is due to different types of APC in the primary culture; when OVA-specific T cells are cultured with various amounts of OVA peptide and irradiated splenocytes, a pattern of T helper subset development occurs, similar to what is observed for Ac1–11[4Y] (N. Hosken and A. O'Garra, personal communication).

The differential effects both in vivo and in vitro of these three peptides probably stems from their different affinities for I-A^u. The half-lives of each of the peptide/MHC complexes varies tremendously, and as a consequence even at very high concentrations of Ac1–11 the number of Ac1–11/I-A^u complexes never reaches the number of complexes of Ac1–11[4Y]/I-A^u at moderate concentrations of Ac1–

◀──

Fig. 1A,B. Cytokine profiles of splenocytes from mice injected with phosphate-buffered saline (PBS), Ac1–11, Ac1–11[4A], or Ac1–11[4Y] 1 day after administration. Increasing amounts of Th2 type cytokines, IL-4 and IL-10, were detected, while decreasing amounts of the Th1 type cytokine interferon (IFN-γ) were detected, depending on the peptide injected. Cells were cultured at 3–5×10^5 cells per well in 96-well round bottom plates with either medium only, Ac1–11, Ac1–11[4A], or Ac1–11[4Y]. Supernatants were collected 45–52 h later and the amount of IL-2, IFN-γ, IL-4, and IL-10 was determined by ELISA. **A** Cytokine profiles of splenocytes from mice injected with PBS or 2.4 mg Ac1–11, Ac1–11[4A], or Ac1–11[4Y] 1 day after administration. **B** Cytokine profiles of splenocytes from mice injected with PBS or 2.4 mg Ac1–11, Ac1–11[4A], or Ac1–11[4Y] 6 days after administration. Similar to day 1, increasing amounts of Th2 type cytokines, IL-4 and IL-10, were detected, while decreasing amounts of the Th1 type cytokines, IL-2 and IFN-γ, were detected, depending on the peptide injected. No cytokines were detected in supernatants from cells incubated with medium only. Reproduced from the Journal of Experimental Medicine, 1997, 185:583–600, by copyright permission of the Rockefeller University Press

Fig. 2. The induction of Th1 and Th2 type cells in vitro. Lymph node cells from a myelin basic protein (MBP)-specific T cell receptor (TCR) transgenic mouse were stained with CD8, B220, Mac-1, CD69, and CD44, and the negative cells were collected by flow cytometry as naive, CD4 + T cells (reanalysis of cells by flow cytometry indicated a population >98% CD4+). These cells were cultured in vitro at a concentration of 10^5 CD4+ cells and 10^6 irradiated non-transgenic syngeneic splenocytes per well in 24-well plates with varying concentrations of Ac1 11, Ac1 11[4A], or Ac1 11[4Y]. Ten days later, cells were washed several times and restimulated at a concentration of 5×10^4 T cells plus 3×10^5 irradiated splenocytes per well in 96-well round bottom plates with 10 μM Ac1 11, Ac1 11[4A], or Ac1 11[4Y]. Supernatants were collected 48 h and cytokines were detected by ELISA. In general, Ac1 11 induced Th1 type responses regardless of primary concentration, Ac1 11[4A] induced Th2 type responses at high concentrations, and Th1 type responses at low concentrations, and Ac1 11[4Y] induced primarily Th2 type responses, except at the lowest two doses; nd, not determined. Previous experiments indicated that too few live cells were recovered at the doses for which cytokine levels were not determined to test a secondary response. Reproduced from the Journal of Experimental Medicine, 1997, 185:583 600, by copyright permission of the Rockefeller University Press

11[4Y]. Although position four in Ac1 11 is thought to interact only with the MHC, the TCR affinities for Ac1–11, Ac1–11[4A] or Ac1–11[4Y] complexed to I-A^u are unknown. If they do differ significantly, these peptides could be considered altered peptide ligands (APLs), or analogues of the wild-type peptide that induce a partial or differential T cell response (SLOAN-LANCASTER and ALLEN 1996), and have a lower affinity, when complexed to the MHC, for the TCR (LYONS et al. 1996; ALAM et al. 1996).

13 Altered Peptide Ligands Can Induce Th2 Responses as Therapy for Experimental Autoimmune Encephalomyelitis

Altered peptide ligands have been shown previously to induce different T helper subsets. IL-4 mRNA was detectable after immunization with APLs of human collagen IV peptide but not after the wild type peptide (PFEIFFER et al. 1995). Several studies have shown that antagonists (which are peptides that putatively compete for binding to the TCR but do not activate the T cell) of PLP 139 151 are efficient inhibitors of EAE (FRANCO et al. 1994; KUCHROO et al. 1994; NICHOLSON et al. 1995). Immunization with a particular antagonist, Q144, induces T cells that are cross-reactive with the native PLP 139 151 peptide (W144) and are either Th0 or Th2 in phenotype. Furthermore, cells elicited by Q144 and stimulated in vitro with Q144 offered mild protection, whereas cells stimulated with W144 offered the most protection against EAE, emphasizing the importance of the cross-reactive T cells in protection. Normalizing for the proliferative response, there was no difference in cytokine production with either W144 or Q144, but T cell lines derived from Q144 immunization produced significantly more IL-10 than those derived from W144 immunization when stimulated in vitro with either peptide (NICHOLSON et al. 1995).

Similar studies have been done using antagonists of MBP peptide 87 99, which induces EAE in Lewis rats, in which one antagonist could inhibit EAE when co-immunized with the wild type peptide (KARIN et al. 1994). In addition, lymph node

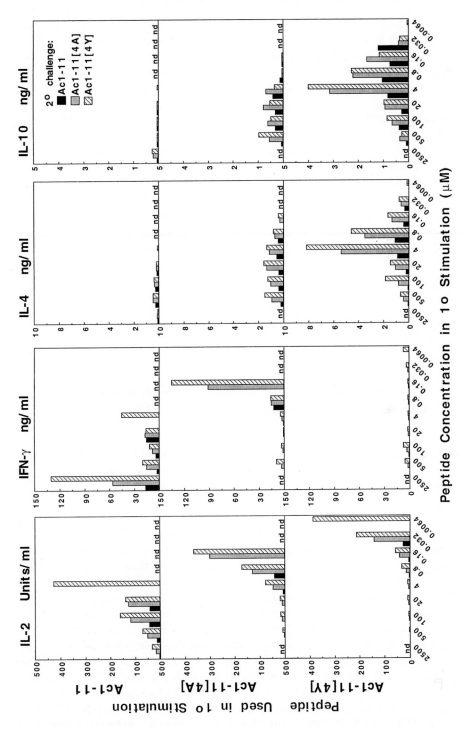

cells produced less IFN-γ and TNF-α after coimmunization with the antagonist plus the 87–99 peptide than after immunization with the 87–99 peptide alone (KARIN et al. 1994). Furthermore, this same antagonist could ameliorate EAE, although the mechanism of such therapy was not attributed to a switch in Th1 to Th2 responses. In a related study, a different APL of 87–99, 87–99 (96P→A), was able to prevent or treat paralysis induced in (PL/JxSJL/J)F1 mice with an encephalitogenic T cell line (BROCKE et al. 1996). This particular peptide did not influence the in vitro proliferative response of the T cell line, and so probably does not exert its effect through competition with either the MHC or TCR. Although the initial response to autoantigen in EAE is thought to be restricted to a particular TCR repertoire in B10.PL and PL/J mice, inflammatory infiltrates in the lesions in the central nervous system contain a heterogeneous population of T cells bearing various TCR (BELL et al. 1993). Treatment with the APL resulted in a reduction in the number of Vβ transcripts in the brain accompanied by an absence of transcripts of TNF-α normally seen in EAE (BROCKE et al. 1996). These results indicated that the APL is influencing cytokine secretion by the T cell line. Furthermore, treatment of mice with anti-IL-4 antibody abrogated any beneficial effect of the APL. TNF-α has been shown to be associated with occurrence of MS (RAINE 1995), and IL-4 itself is associated with recovery from disease and is seen in MS lesions (RACKE et al. 1994; CANNELLA and RAINE 1995).

14 Specific Antigen Therapy in Insulin-Dependent Diabetes Mellitus Deviates Response from Th1 to Th2

Several studies have shown that specific antigen therapy in NOD mice induces Th2 responses associated with prevention or delay of diabetes. Treatment or prevention of diabetes has in the past utilized drugs that generally suppress the immune system, or block T cell or APC function indiscriminately. These approaches may have deleterious side effects, and thus specific antigen therapy remains an attractive alternative. Unlike EAE or MS, however, the autoantigens in diabetes have not been clearly identified, although a number of proteins have been identified that are thought to be involved in the diabetogenic process. Of these, glutamic acid decarboxylase 65 (GAD65) is thought to play a critical role in initiating disease (BAEKKESKOV et al. 1990), as prediabetic patients and NOD mice have antibodies against GAD65 present in their sera (HAGOPIAN et al. 1993; TISCH et al. 1993; KAUFMAN et al. 1993). The importance of GAD65 in diabetes was indicated when protein or peptides derived therefrom given intrathymically, intranasally, or intravenously to female NOD mice prior to diabetic symptoms prevents insulitis and diabetes (TISCH et al. 1993, KAUFMAN et al. 1993). The mechanism of such therapy may lie in a shift from a Th1 to Th2 response, as reduction of the incidence of diabetes by intranasal administration of a mixture of peptides from GAD65 to 2–3-week-old female NOD mice correlated with an increase in IgG1 responses to

GAD65 and IL-5 production by T cells, and a decrease in IFN-γ production (TIAN et al. 1996a). Furthermore, this protection was associated with the CD4+ T cell population, as transfer of these cells from peptide-treated mice were able to protect NOD-scid/scid mice from adoptive transfer-induced diabetes (TIAN et al. 1996a). Immune deviation by treatment of prediabetic mice with a single injection of GAD65 in IFA intraperitoneally reduced the incidence of diabetes and protected syngeneic islet grafts from rejection as well (TIAN et al. 1996b).

In an another study, immunization with GAD65 peptide 524 543 prior to injection of cyclophosphamide, which accelerates diabetes onset, in 4-week-old female NOD mice delayed the onset and reduced the incidence of diabetes. This protection was correlated with a strong Th2 response to GAD peptide 524 543 up to 3 months following immunization (SA et al. 1996). Furthermore, T cells from immunized mice could inhibit adoptively transferred diabetes development (SA et al. 1996). Since some experiments have not been able to confirm that 524 543 peptide is an immunodominant peptide in the spontaneous response to GAD65 (CHEN et al. 1994; SA et al. 1996), this experiment may be initiating a Th2 response to 524 543 that down-regulates spontaneous responses to other autoantigens involved in diabetes. RAMIYA et al. 1996, however, were unable to show any beneficial effect of giving GAD65 524 543 in IFA.

Because most treatment of diabetes would involve treatment of patients in which extensive β-cell destruction has already taken place, a more relevant study showed that four intravenous treatments of whole GAD65 given to 12-week-old female NOD mice, in which overt insulitis has developed, prevented diabetes (Tisch et al. 1994, 1998). Protection from diabetes directly correlated with reduced T cell proliferation to GAD65 and a panel of other autoantigens. Furthermore, CD4+ T cells from GAD65-treated mice were able to suppress adoptive transfer of diabetes, suggesting that intravenous GAD65 induced a regulatory CD4+ T cell population, consistent with a shift in the response from Th1 to Th2. Most of these studies show that T cell proliferative responses to the injected protein or peptide are markedly reduced, and there is an increase in Th2 type responses in the reduced proliferative response that remains. These findings suggest that T cells are being activated initially and differentiating into Th2 cells and/or dying through apoptosis. Those that survive are Th2 types, which antagonize remaining or new Th1 responses to autoantigen. Interpretation of these results parallels those results in the MBP-specific TCR transgenic model discussed above.

While GAD-specific responses clearly possess key roles in diabetes, other autoantigens may be critical players in diabetes. Antibodies to insulin are found in diabetic patients, and while antibody responses to insulin do not predict incidence of IDDM as well as anti-GAD antibodies, a high frequency of insulin-specific T cells has been found in pancreatic islets of NOD mice (WEGMANN et al. 1994). Thus, some studies have focused on the efficacy of insulin treatment. Insulin or the insulin B chain given subcutaneously in incomplete Freund's adjuvant protects mice from diabetes and correlates with reduced production of IFN-γ in islets (MUIR et al. 1995). Further studies using alum as an adjuvant show that insulin B chain and diphtheria tetanus toxoid-acellular pertussis vaccine (DTP) significantly reduced the incidence of diabetes from 100% to 27% (RAMIYA et al. 1996). Immune deviation from a Th1 to a Th2 response was proposed as the mechanism, as intra-islet cytokine pattern in DTP and insulin B chain immunized mice exhibited higher levels of IL-4, IL-10, and TGF-β, and a higher level of Th2-induced IgG1 antibodies relative to untreated, DTP- or DTP + insulin A chain immunized mice (RAMIYA et al. 1996). A single subcutaneous injection of insulin B chain peptide 9–23 in IFA, or multiple intranasal treatments with peptide, decreases the incidence of diabetes; moreover, lymph node cells from mice treated intranasally produced predominantly Th2 type cytokines (DANIEL and WEGMANN 1996).

Studies have shown that immunization of NOD mice with CFA protects mice from diabetes (QIN et al. 1993). This is in part due to the up-regulation of TNF production by CFA (RABINOVITCH et al. 1994). Further study indicated that T cells could be isolated from NOD mice that react to a 65 kDa HSP from *Mycobacterium*, one of the components of CFA, and that, depending on the type of immunization, diabetes could be induced or prevented in NOD mice with whole HSP or a peptide derived therefrom (ELIAS et al. 1990, 1991). Ongoing disease in NOD mice could be treated with a peptide derived from the human heat shock protein (hsp60), which is homologous to the mycobacterial HSP (ELIAS and COHEN 1995, 1996). T cell responses to hsp60 were reduced in treated mice, and transfer of cells from peptide-treated mice could suppress adoptive transfer of disease (ELIAS et al. 1991). These findings are consistent with a switch from a Th1 response to a Th2 response, although no direct evidence was provided in these studies. T cell reactivity to p277 can be detected in C57BL/KsJ mice in which diabetes has been induced by low doses of streptozotocin. Two doses of 100 micrograms of p277 peptide given in mineral oil were sufficient to prevent hyperglycemia. Moreover, this prevention was accompanied by a lack of T cell proliferation to p277 and an up-regulation of anti-p277 antibodies of the IgG1 and IgG2b isotypes that are associated with Th2 responses (ELIAS and COHEN 1996). Another peptide spanning the amino acids 509–528 of the GAD65 protein, was unable to protect from diabetes, and induced peptide-specific antibodies primarily of the IgG2a isotype.

In summary, antigen-specific therapies hold much promise in the therapy of autoimmune disease, as indicated by the evidence presented above. One recent study, however, suggests that antigen-specific therapy may not always result in a protective response, and that the distinction between Th1 and Th2 responses is not so clear. EAE induced by rat myelin oligodendrocyte glycoprotein (MOG) in marmosets can be delayed when marmosets are treated six times every other day starting at day 7 following induction with soluble MOG given intraperitoneally. Once treatment ceases, however, an acute, lethal form of EAE that is more severe than that found in controls develops. The development of the lethal EAE correlated with the appearance of T cell proliferative response to MOG and an increase in MOG-specific antibodies four to eight times higher than that seen in the controls. Increased synthesis of IL-10 and IL-6 mRNA and decreased synthesis of IFN-γ and TNF-α mRNA was observed, indicating that a shift from Th1 to Th2 response in lymph node cells and peripheral blood mononuclear cells had occurred (GENAIN et al. 1996). Thus, the shift to a Th2 response may in fact have enhanced production of pathogenic autoantibodies, which in this model of EAE induce demyelination. The relevance of this particular study remains unclear, however. Treatment with whole protein is more likely to elicit an antibody response rather than treatment with peptide. Antibody responses enhance MOG-induced EAE but not MBP- or PLP-induced EAE. In addition, although myelin-specific antibodies are found in MS patients (PANITCH et al. 1980; SCOLDING et al. 1989), the role of the antibody response in tissue damage is thought to be minimal. Tables 2 and 3 summarize the effects of specific antigen therapy on EAE and IDDM.

15 Oral Tolerance as Specific Antigen Therapy

Oral administration of antigens has been used to induce tolerance in autoimmune diseases (WEINER et al. 1994). A number of mechanisms have been proposed, including active suppression and T cell anergy (WHITACRE et al. 1991; FRIEDMAN and WEINER 1994). More recently, oral administration of MBP to SJL/J mice was shown to induce antigen-specific T cells in the Peyer's patches in the intestinal tract that secrete primarily IL-4, IL-10, and/or TGF-β1 (CHEN et al. 1994). These findings were extended to show that moderate doses of ovalbumin given orally to mice transgenic for an ovalbumin-specific TCR induce antigen-specific cells that secrete IL-4, IL-10, and/or TGF-β1 (CHEN et al. 1995). Similar findings were reported in mice transgenic for a MBP-specific TCR (CHEN et al. 1996). At higher doses, deletion of ovalbumin-specific cells is observed in both Th1 and Th2 compartments (CHEN et al. 1995). Whole myelin has been used successfully to treat chronic relapsing EAE in SJL/J mice (AL-SABBAGH et al. 1996), and MS patients had fewer relapses when fed whole myelin (WEINER et al. 1993), although more data need to be gathered to make a definitive conclusion. Circulating lymphocytes from myelin-fed MS patients were found to have an increased frequency of TGF-β1-producing cells specific for either MBP or PLP (FUKAURA et al. 1996). In NOD mice, feeding of insulin suppressed insulitis and was correlated with an increase in expression of IL-4, IL-10, TGF-β1, and prostaglandin E within the mononuclear cell infiltrate in the pancreatic islets (HANCOCK et al. 1995). CD4+ T cells isolated from pancreatic islets were found to secrete IL-10, IFN-γ and TGF-β, and suppress diabetes, when adoptively transferred into NOD mice. The suppressive activity depended on TGF-β, as only antibodies to TGF-β given in vivo after adoptive transfer of the T cells could counteract suppression of disease (HAN et al. 1996).

    In another study, oral administration of 20 mg of ovalbumin protein to normal mice induced OVA-specific CD8+ cytotoxic lymphocytes (CTLs). Mice that were transgenic for ovalbumin expressed under the rat insulin promoter, and chimeric at a 1:4 ratio for transgenic T cells expressing a class I-restricted TCR specific for OVA, did not get spontaneous disease, but when fed one dose of 60 mg OVA almost 50% of the mice became diabetic (BLANAS et al. 1996). These findings indicate the importance of CD8+ T cells in the diabetic immune response, but also

Table 2. Specific antigen therapy: experimental autoimmune encephalomyelitis (EAE)

Antigen	Route	Effect	Mechanism	Reference
MBP Ac1–11 Ac1–11[4A]	s.c. in IFA	Prevents, treats	ND	Smilek et al. 1991
Ac1–11, 35–47	i.p. in IFA	Prevents, treats	Anergy	Gaur et al. 1992
Ac1–11, Ac1–11[4Y]	Intranasally in PBS	Prevents	ND	Metzler and Wraith 1993
Ac1–11, MBP	i.v. in PBS	Prevents	Deletion	Critchfield et al. 1994
Ac1–11, Ac1–11[4Y]	i.p. in PBS	Prevents, treats	ND	Liu and Wraith 1996
Ac1–11, Ac1–11[4Y]	i.v. in PBS	Treats	ND	Samson and Smilek 1995
PLP 139–151 APL	Coinjection in CFA	Prevents	ND	Kuchroo et al. 1994
	i.p. in PBS	Treats		
PLP 139–151 APLs	Coinjection in CFA	Prevents	ND	Franco et al. 1994
PLP 139–151 APL	Coinjection in CFA	Prevents	Th1→Th2	Nicholson et al. 1995
	i.p. in PBS	Treats	ND	
MBP 87–99 APL	Coinjection in CFA	Prevents	Th1 down-regulated	Karin et al. 1994
	i.p. in PBS	Treats		
MBP 87–99 APL	i.p. in PBS	Treats	Th1→Th2	Brocke et al. 1996
MOG	i.p. in PBS	Delays onset, then exacerbates	Th1→Th2	Genain et al. 1996

PBS, phosphate-buffered saline; CFA, complete Freund's adjuvant; MBP, myelin basic protein; PLP, proteolipid protein; APL, altered peptide ligands; MOG, myelin oligodendrocyte glycoprotein; ND, not determined

Table 3. Specific antigen therapy: diabetes in NOD mice

Antigen	Route	Effect	Mechanism	Reference
GAD65	One dose i.t. at 3 weeks old	Reduces incidence	ND	Tisch et al. 1993, 1994
GAD65	One dose i.v. at 3 weeks old	Reduces incidence	ND	Kaufman et al. 1993
GAD65 peptides	One dose intranasally at 2–3 weeks old	Reduces incidence	Th1→Th2	Tian et al. 1996a
GAD65	One dose in IFA i.p. at 8 weeks old	Reduces incidence	Th1→Th2	Tian et al. 1996b
GAD65	Four doses i.v. to 12-week-old NOD mice	Reduces incidence	Th1→Th2	Tisch et al. 1994, 1998
GAD65 p 524–543	One dose at 3 or 7 weeks old in IFA Diabetes induced with CY at 10 weeks old	Reduces incidence	Th1→Th2	Saï et al. 1996
Insulin, insulin B chain	Four doses in IFA at 4 weeks old every 4–8 days	Reduces incidence	Th1→Th2	Muir et al. 1995
Insulin B chain	One dose in alum at 3 weeks old	Reduces incidence	Th1→Th2	Ramiya et al. 1996
GAD65 peptides	One dose in alum at 3 weeks old	No effect	NA	
Insulin B p9–23	One dose in IFA at 3 weeks old	Reduces incidence		
Insulin B + DTP	One dose in alum at 3 weeks old	Reduces incidence to 27%		
Insulin B p9–23	One dose in IFA at 4 weeks old Three daily doses at 4 weeks old every 4–5 weeks in PBS intranasally	Reduces incidence Reduces incidence	ND Th1→Th2	Daniel and Wegmann 1996

Table 3 (*cont.*)

Antigen	Route	Effect	Mechanism	Reference
HSP 65 protein	One dose i.p. in IFA or PBS at 4–5 weeks old	Reduces incidence	ND	ELIAS et al. 1990
HSP peptide p 277	One dose i.p. in IFA at 4 weeks old	Reduces incidence	ND	ELIAS et al. 1991
HSP peptide p 277	One dose i.p. in IFA at 7, 12, 15, or 17 weeks old	Reduces incidence: most effective at younger ages	ND	ELIAS and COHEN 1995
HSP peptide p 277	One dose 7 days and one dose 85 days after streptozotocin in IFA	Reduces incidence of streptozocin-induced IDDM	Th1→Th2	ELIAS and COHEN 1996

IFA incomplete Freund's adjuvant; IDDM, insulin-dependent diabetes mellitus; HSP, heat shock protein; PBS, phosphate-buffered saline; CY, cyclophosphamide; NOD, nonobese diabetic mouse; ND, not done.

suggest that (at least oral) administration of specific antigen may induce deleterious responses in one arm of an autoimmune response. These results conflict with oral administration of insulin, which delays onset and decreases the incidence of diabetes in NOD mice, and which is correlated with a Th2 response to insulin. One study did find that insulin feeding induced CD8+ T cells that could enhance disease after adoptive transfer (BERGEROT et al. 1994); however, in the study by BLANAS et al. (1996), the Th phenotype of the CD4+ T cell compartment was not characterized. It is possible that the CD4+ T cells induced at such a high dose (60 mg) of ovalbumin differentiated into Th1 type cells, rather than Th2, and were unable to suppress CTL activity. Use of a large amount of whole ovalbumin, rather than peptide, may have allowed antigen to go to lymph nodes outside of the intestinal tract, where Th1 responses are more likely to be elicited.

16 Costimulatory Molecules as Targets for Therapy

Recently, costimulatory interactions between B7-1 and B7-2 and CD28 or CTLA4 needed for full activation of a T cell have been shown to influence the T helper subset outcome. Transfectants expressing B7-1 or B7-2 preferentially induced Th1 or Th2 type responses, respectively (FREEMAN et al. 1995; RANGER et al. 1996). Approximately at the same time, studies in EAE showed that blocking B7-1 allowed Th2 responses to develop following immunization with PLP 139–151, protecting the mice from EAE, whereas blocking B7-2 enhanced responses (KUCHROO et al. 1995). Preferential up-regulation of B7-1 occurs during EAE, and in agreement with KUCHROO et al. 1995, blocking B7-1 with anti-B7-1 Fab fragments blocked EAE (MILLER et al. 1995).

In contrast, antibodies to B7-1 or B7-2 given to NOD mice either accelerated or delayed onset of diabetes, respectively (LENSCHOW et al. 1995). The difference between this study and others in EAE may be due to the ability of the different antibodies used in these studies to block or stimulate through B7. Diabetes is accelerated in CD28 deficient mice, however; this earlier onset is associated with the production of more IL-2 and IFN-γ and less IL-4 by GAD-specific cells, suggesting that a Th2 response is dependent on CD28 signaling (LENSCHOW et al. 1996). Expression of B7-1 or TNF-α alone on pancreatic β-cells is not sufficient to induce diabetes, while coexpression of both does induce diabetes in normally nonsusceptible strains of mice (HERRERA et al. 1994; GUERDER et al. 1994). B7-1 expression alone on β-cells on the NOD background, however, is sufficient to accelerate diabetes onset (WONG et al. 1995). When B7-1 and the glycoprotein from lymphocytic choriomeningitis virus are coexpressed on pancreatic islet cells in a non-susceptible strain, spontaneous IDDM occurs. In situ staining for Th1 and Th2 type cytokines in the pancreas revealed that B7-1 expression enhances the Th1 phenotype (VON HERRATH et al. 1995). Studies that examined B7 expression in the pancreas found that overt diabetes, but not insulitis, was strongly correlated with both B7-1 and B7-2 expression (STEPHENS and KAY 1995).

The strong effect of costimulation is thought not only to deliver a necessary second signal (apart from the TCR/peptide/MHC interaction) for T cell activation, but also to provide a stronger signal that ultimately influences T helper differentiation. While many studies suggest that the different B7 molecules play a different role in T helper subset differentiation, their roles may ultimately be in providing different strengths of signal, based on the expression and affinity for CD28 for B7-1 and B7-2. Increasing the strength of signal solely through CD28 induces an increase in the production of IL-5 (RULIFSON et al. 1997). These experiments are paralleled by those that examine the effect of dose of antigen on T helper development, which found that increasing doses of antigen leads to T helper 2 responses (HOSKEN et al. 1995; PEARSON et al. 1997; BRETSCHER et al. 1992). In contrast, exclusive signals through CTLA4 down-regulate T cell activation, possibly through inhibition of the cell cycle progression from G1 to S (KRUMMEL and ALLISON 1996). Thus, the combination of inhibition of CD28-induced activation signals plus specific peptide therapy may result in the most effective blockade of an autoimmune response.

17 Expression of Class II MHC Molecules Protects from Diabetes

Class II MHC transgenic mice on the NOD background are protected from diabetes (NISHIMOTO et al. 1987; LUND et al. 1990; SLATTERY et al. 1990; SINGER et al. 1993; HANSON et al. 1996; WHERRETT et al. 1997). While the mechanism is unclear, several themes have emerged. Studies on class II MHC transgenic mice on the NOD and other backgrounds reveal that overexpression sometimes leads to a variable B cell defect that is associated with enhanced Th2 responses that are dependent upon IL-4, IL-5, and possibly IL-6 expression (SINGER et al. 1996). This generalized skewing toward a Th2 response would then protect mice from diabetes by preventing autoreactive Th1 responses.

In class II MHC transgenic NOD mice which are free of this B cell defect, expression of IDDM resistant MHC class II alleles results in a decreased incidence of diabetes (WHERRETT et al. 1997). Resistant class II MHC molecules may present either the same peptide derived from a β-cell antigen, or a different β-cell antigenic peptide altogether; in either case, presentation of such a peptide/MHC complex to potentially diabetogenic T cells may initiate protective Th2 responses, rather than deleterious Th1 responses. The NOD class II molecule, I-A^{g7}, is the least stable of the different I-A and I-E haplotypes, suggesting that a different class II MHC molecule, even presenting the same autoantigenic peptide as I-A^{g7}, may elicit Th2 responses, as it would be more stable and hence deliver stronger signals to the responding T cells. Another mechanism of protection could come from a dose effect by increasing the number of MHC molecules present on APC, thereby increasing the signal and thus skewing immune responses to Th2. The exact mechanisms of protection through transgenic expression of resistant class II MHC molecules in the NOD mouse remain to be identified. The potential treatment of diabetes through anti-class II MHC gene therapy, however, remains a viable option.

18 Neonatal Tolerance

Neonatal injection of antigen either whole protein or peptide has resulted in tolerance to subsequent challenge with the same antigen (NOSSAL 1983). This type of immunization has been used to inhibit EAE, CIA, and diabetes (CLAYTON et al. 1989; MYERS et al. 1989; PETERSEN et al. 1994). Neonatal tolerance is thought to induce clonal anergy and/or deletion, but recently was also found to induce Th2 type responses when encountering peptide in a secondary challenge (SINGH et al. 1996; FORSTHUBER et al. 1996; RIDGE et al. 1996; SARZOTTI et al. 1996). Thus the interpretation is that neonates do not necessarily have a different type of immune response; rather, the fewer number of cells in a neonate compared to that of an adult would receive a stronger signal during challenge with the same amount of antigen. The consequences of neonatal tolerance induction (and immune deviation from Th1 to Th2 in general) are that antibody-mediated diseases might be exacerbated, while Th1-mediated autoimmune disease could be ameliorated (SINGH et al. 1996).

19 Conclusions

Clearly, Th1-mediated autoimmune diseases can be ameliorated by diverting responses toward a Th2 phenotype. Many groups are now searching for the most efficient methods for deviating the immune response from a deleterious Th1 to a beneficial Th2 response. Two main approaches have been taken. The first is direct treatment with cytokines that dominate the response and direct it toward a Th2 phenotype, such as IL-4 and IL-10, or antibodies and antagonists to Th1-directing cytokines such as IL-12, or those thought to be more directly involved in causing tissue damage, such as IL-1 and TNF-α. However, timing and route of administration can determine whether such therapy exacerbates or improves symptoms. IL-4 and IL-10 for example, give the most benefit when given early on and systemically, rather than locally. This type of therapy may prove difficult to apply in chronic progressive forms of disease, as in the case of MS, where anti-TNF antibody does not appear to have any beneficial effect, or in treating diabetes before complete β-cell destruction.

The second approach of using antigen-specific therapy has yielded promising results as well. Several recent studies have indicated that soluble antigen therapy preferentially induces specific Th2 responses that protect against autoimmunity. These strategies may prove to be more beneficial in chronic disease such as RA and MS as successful treatment of at least EAE can occur during the symptomatic stage of disease. In diabetes, the most effective time of treatment with specific antigen would be prior to β-cell destruction. Thus, development of reliable diagnostic techniques to identify those people who are likely to become diabetic are of paramount importance. The use of cytokines can be dangerous, since studies in animal models result in paradoxical effects and unwanted side effects. Moreover, while antibodies to cytokines are beneficial initially, they become less effective, since an immune response to the anti-cytokine antibodies renders them useless. Current techniques, such as the replacement of the murine antibody germline genes with those of human may allow the development of human antibodies that are non-immunogenic. Specific peptides or even small molecules which block the TCR have the advantage in that an unwanted immune response is less likely to occur during treatment. Perhaps the approach best taken is combinatorial therapy, in which brief treatment with Th2 cytokines, antibodies to Th1 type cytokines such as IL-12, and specific peptides or small molecule blockers of the TCR are given in combination at the same time, or at various stages of disease.

References

Acha Orbea H, Mitchell D, Timmerman L, Wraith D, Tausch G, Waldor M, Zamvil S, McDevitt HO, Steinman L (1988) Limited heterogeneity of T cell receptors from lymphocytes mediating autoimmune encephalomyelitis allows specific immune intervention. Cell 54:263 273

al-Janadi N, al-Dalaan A, al-Balla S, Raziuddin S (1996) CD4 + T cell inducible immunoregulatory cytokine response in rheumatoid arthritis. J Rheumatol 23:809 814

al-Sabbagh AM, Goad EP, Weiner HL, Nelson PA (1996) Decreased CNS inflammation and absence of clinical exacerbation of disease after six months oral administration of bovine myelin in diseased SJL/J mice with chronic relapsing experimental autoimmune encephalomyelitis. J Neurosci Res 45:424 429

Alam SM, Travers PJ, Wung JL, Nasholds W, Redpath S, Jameson SC, Gascoigne NR (1996) T-cell-receptor affinity and thymocyte positive selection. Nature 381:616 620

Baekkeskov S, Aanstoot HJ, Christgau S, Reetz A, Solimena M, Cascalho M, Folli F, Richter-Olesen H, DeCamilli P, Camilli PD (1990) Identification of the 64 K autoantigen in insulin-dependent diabetes as the GABA-synthesizing enzyme glutamic acid decarboxylase. Nature 347:151 156

Baker D, Rosenwasser OA, O'Neill JK, Turk JL (1995) Genetic analysis of experimental allergic encephalomyelitis in mice. J Immunol 155:4046 4051

Balashov KE, Smith DR, Khoury SM, Hafler DA, Weiner HL (1997) Increased interleukin 12 production in progressive multiple sclerosis: Induction by activated CD4 + T cells via CD40 ligand. Proc Natl Acad Sci USA 94:599 603

Baron JL, Madri JA, Ruddle NH, Hashim G, Janeway CA Jr (1993) Surface expression of alpha 4 integrin by CD4 T cells is required for their entry into brain parenchyma. J Exp Med 177:57 68

Bell RB, Lindsey JW, Sobel RA, Hodgkinson S, Steinman L (1993) Diverse T cell receptor V beta gene usage in the central nervous system in experimental allergic encephalomyelitis. J Immunol 150:4085 4092

Bergerot I, Fabien N, Maguer V, Thivolet C (1994) Oral administration of human insulin to NOD mice generates CD4 + T cells that suppress adoptive transfer of diabetes. J Autoimmunity 7:655 663

Bessis N, Boissier MC, Ferrara P, Blankenstein T, Fradelizi D, Fournier C (1996) Attenuation of collagen-induced arthritis in mice by treatment with vector cells engineered to secrete interleukin-13. Eur J Immunol 26:2399 2403

Billiau A, Heremans H, Vandekerckhove F, Dijkmans R, Sobis H, Meulepas E, Carton H (1988) Enhancement of experimental allergic encephalomyelitis in mice by antibodies against IFN-gamma. J Immunol 140(5):1506 1510

Birnbaum G, Kotilinek L, Schlievert P, Clark HB, Trotter J, Horvath E, Gao E, Cox M, Braun PE (1996) Heat shock proteins and experimental autoimmune encephalomyelitis (EAE): I. Immunization with a peptide of the myelin protein 2′,3′ cyclic nucleotide 3′ phosphodiesterase that is cross-reactive with a heat shock protein alters the course of EAE. J Neurosci Res 44:381 396

Blanas E, Carbone FR, Allison J, Miller JFAP, Heath WR (1996) Induction of autoimmune diabetes by oral administration of autoantigen. Science 274:1707 1709

Bliss J, Van Cleave V, Murray K, Wiencis A, Ketchum M, Maylor R, Haire T, Resmini C, Abbas AK, Wolf SF (1996) IL-12, as an adjuvant, promotes a T helper 1 cell, but does not suppress a T helper 2 cell recall response. J Immunol 156:887 894

Bretscher PA, Wei G, Menon JN, Bielefeldt-Ohmann H (1992) Establishment of stable, cell-mediated immunity that makes "susceptible" mice resistant to Leishmania major. Science 257:539 542

Brocke S, Gijbels K, Allegretta M, Ferber I, Piercy C, Blankenstein T, Martin R, Utz U, Karin N, Mitchell D, et al. (1996) Treatment of experimental encephalomyelitis with a peptide analogue of myelin basic protein. Nature 379:343 346

Brod SA, Marshall GD Jr, Henninger EM, Sriram S, Khan M, Wolinsky JS (1996) Interferon-beta 1b treatment decreases tumor necrosis factor-alpha and increases interleukin-6 production in multiple sclerosis. Neurology 46:1633 1638

Burstein HJ, Abbas AK (1993) In vivo role of interleukin 4 in T cell tolerance induced by aqueous protein antigen. J Exp Med 177:457 463

Burstein HJ, Shea CM, Abbas AK (1992) Aqueous antigens induce in vivo tolerance selectively in IL-2- and IFN-γ-producing (Th1) cells. J Immunol 148:3687 3691

Campbell IL, Kay TW, Oxbrow L, Harrison LC (1991) Essential role for interferon-gamma and interleukin-6 in autoimmune insulin-dependent diabetes in NOD/Wehi mice. J Clin Invest 87:739 742

Cannella B, Raine CS (1995) The adhesion molecule and cytokine profile of multiple sclerosis lesions. Ann Neurol 37:424 435

Cash E, Minty A, Ferrara P, Caput D, Fradelizi D, Rott O (1994) Macrophage-inactivating IL-13 suppresses experimental autoimmune encephalomyelitis in rats. J Immunol 153:4258 4267

Chen Y, Inobe J, Kuchroo VK, Baron JL, Janeway CA Jr, Weiner HL (1996) Oral tolerance in myelin basic protein T-cell receptor transgenic mice: suppression of autoimmune encephalomyelitis and dose-dependent induction of regulatory cells. Proc Natl Acad Sci USA 93:388 391

Chen Y, Inobe J, Marks R, Gonnella P, Kuchroo VK, Weiner HL (1995) Peripheral deletion of antigen-reactive T cells in oral tolerance. Nature 376:177 180

Chen Y, Kuchroo VK, Inobe J, Hafler DA, Weiner HL (1994) Regulatory T cell clones induced by oral tolerance: suppression of autoimmune encephalomyelitis. Science 265:1237 1240

Chen SL, Whiteley PJ, Freed DC, Rothbard JB, Peterson LB, Wicker LS (1994) Responses of NOD congenic mice to a glutamic acid decarboxylase-derived peptide. J Autoimmun 7:635 641

Chu CQ, Londei M (1996) Induction of Th2 cytokines and control of collagen-induced arthritis by nondepleting anti-CD4 Abs. J Immunol 157:2685 2689

Clayton JP, Gammon G, Ando DG, Kono D, Hood L, Sercarz EE (1989) Peptide-specific prevention of experimental allergic encephalomyelitis: neonatal tolerance induced to the dominant T cell determinant of myelin basic protein. J Exp Med 169:1681 1691

Cohen SB, Katsikis PD, Chu CQ, Thomssen H, Webb LM, Maini RN, Londei M, Feldmann M (1995) High level of interleukin-10 production by the activated T cell population within the rheumatoid synovial membrane. Arthr Rheum 38:946 952

Conboy IM, Dekruyff RH, Tate KM, Cao ZA, Moore TA, Umetsu DT, Jones PP (1997) Novel genetic regulation of T helper 1 (Th1)/Th2 cytokine production and encephalitogenicity in inbred mouse strains. J Exp Med 185:439 451

Constant S, Pfeiffer C, Woodard A, Pasqualini T, Bottomly K (1995) Extent of T cell receptor ligation can determine the function differentiation of naive CD4 + T cell. J Exp Med 182:1591 1596

Crisi GM, Santambrogio L, Hochwald GM, Smith SR, Carlino JA, Thorbecke GJ (1995) Staphylococcal enterotoxin B and tumor-necrosis factor-alpha-induced relapses of experimental allergic encephalomyelitis: protection by transforming growth factor-beta and interleukin-10. Eur J Immunol 25:3035 3040

Critchfield JM, Racke MK, Zuniga-Pflucker JC, Cannella B, Raine CS, Goverman J, Lenardo MJ (1994) T cell deletion in high antigen dose therapy of autoimmune encephalomyelitis. Science 263:1139 1143

Cua DJ, Coffman RL, Stohlman SA (1996) Exposure to T helper 2 cytokines in vivo before encounter with antigen selects for T helper subsets via alterations in antigen-presenting cell function. J Immunol 157:2830 2836

Cua DJ, Hinton DR, Stohlman SA (1995a) Self-antigen-induced Th2 responses in experimental allergic encephalomyelitis (EAE)-resistant mice. Th2-mediated suppression of autoimmune disease. J Immunol 155:4052 4059

Cua DJ, Hinton DR, Kirkman L, Stohlman SA (1995b) Macrophages regulate induction of delayed-type hypersensitivity and experimental allergic encephalomyelitis in SJL mice. Eur J Immunol 25:2318 2324

Daniel D, Wegmann DR (1996) Protection of nonobese diabetic mice from diabetes by intranasal or subcutaneous administration of insulin peptide B-(9 23). Proc Natl Acad Sci USA 93:956 960

De Wit D, van Mechelen M, Ryelandt M, Figueiredo AC, Abramowicz A, Goldman M, Bazin H, Urbain J, Leo O (1992) The injection of deaggregated gamma globulins in adult mice induces antigen-specific unresponsiveness of T helper type 1 but not type 2 lymphocytes. J Exp Med 175:9 14

Debray-Sachs M, Carnaud C, Boitard C, Cohen H, Gresser I, Bedossa P, Bach JF (1991) Prevention of diabetes in NOD mice treated with antibody to murine IFN gamma. J Autoimmun 4:237 248

Delespesse G, Yang LP, Shu U, Byun DG, Demeure CE, Ohshima Y, Wu CY, Sarfati M (1996) Role of interleukin-12 in the maturation of naive human CD4 T cells. Ann NY Acad Sci 795:196 201

Duong TT, St. Louis J, Gilbert JJ, Finkelman FD, Strejan GH (1992) Effect of anti-interferon-gamma and anti-interleukin-2 monoclonal antibody treatment on the development of actively and passively induced experimental allergic encephalomyelitis in the SJL/J mouse. J Neuroimmunol 36:105 115

Elias D, Cohen IR (1996) The hsp60 peptide p277 arrests the autoimmune diabetes induced by the toxin streptozotocin. Diabetes 45(9):1168 1172

Elias D, Cohen IR (1995) Treatment of autoimmune diabetes and insulitis in NOD mice with heat shock protein 60 peptide p277. Diabetes 44:1132 1138

Elias D, Markovits D, Reshef T, van der Zee R, Cohen IR (1990) Induction and therapy of autoimmune diabetes in the nonobese diabetic (NOD/Lt) mouse by a 65-kDa heat shock protein. Proc Natl Acad Sci USA 87(4):1576 1580

Elias D, Reshef T, Birk OS, van der Zee R, Walker MD, Cohen IR (1991) Vaccination against autoimmune mouse diabetes with a T-cell epitope of the human 65-kDa heat shock protein. Proc Natl Acad Sci USA 88:3088 3091

Elliott MJ, Mainik RN, Feldmann M, Long-Fox A, Charles P, Katsikis P, Brennan FM, Walker J, Bijl H, Woody JN (1993) Treatment of rheumatoid arthritis with chimeric monoclonal antibodies to Tumor Necrosis factor alpha. Arthr Rheum 36:1681 1690

Faust A, Rothe H, Schade U, Lampeter E, Kolb H (1996) Primary nonfunction of islet grafts in autoimmune diabetic nonobese diabetic mice is prevented by treatment with interleukin-4 and interleukin-10. Transplantation 62:648 652

Feldmann M, Brennan FM, Maini RN (1996) Rheumatoid arthritis. Cell 85:307 310

Ferber IA, Brocke S, Taylor-Edwards C, Ridgway W, Dinisco C, Steinman L, Dalton D, Fathman CG (1996) Mice with a disrupted IFN-gamma gene are susceptible to the induction of experimental autoimmune encephalomyelitis (EAE). J Immunol 156(1):5 7

Forsthuber T, Yip HC, Lehmann PV (1996) Induction of TH1 and TH2 immunity in neonatal mice. Science 271:1728 1730

Franco A, Southwood S, Arrhenius T, Kuchroo VK, Grey HM, Sette A, Ishioka GY (1994) T cell receptor antagonist peptides are highly effective inhibitors of experimental allergic encephalomyelitis. Eur J Immunol 24:940 946

Freeman GJ, Boussiotis VA, Anumanthan A, Bernstein GM, Ke XY, Rennert PD, Gray GS, Gribben JG, Nadler LM (1995) B7-1 and B7-2 do not deliver identical costimulatory signals, since B7-2 but not B7-1 preferentially costimulates the initial production of IL-4. Immunity 2:523 232

Friedman A, Weiner HL (1994) Induction of anergy or active suppression following oral tolerance is determined by antigen dosage. Proc Natl Acad Sci USA 91:6688 6692

Fukaura H, Kent SC, Pietrusewicz MJ, Khoury SJ, Weiner HL, Hafler DA (1996) Induction of circulating myelin basic protein and proteolipid protein-specific transforming growth factor-beta1-secreting Th3 T cells by oral administration of myelin in multiple sclerosis patients. J Clin Invest 98:70 77

Gaur A, Ruberti G, Haspel R, Mayer JP, Fathman CG (1993) Requirement for CD8 + cells in T cell receptor peptide-induced clonal unresponsiveness. Science 259:91 94

Gaur A, Wiers B, Liu A, Rothbard J, Fathman CG (1992) Amelioration of autoimmune encephalomyelitis by myelin basic protein synthetic peptide-induced anergy. Science 258:1491 1494

Genain CP, Abel K, Belmar N, Villinger F, Rosenberg DP, Linington C, Raine CS, Hauser SL (1996) Late complications of immune deviation therapy in a nonhuman primate. Science 274:2054 2057

Gerli R, Muscat C, Bistoni O, Falini B, Tomassini C, Agea E, Tognellini R, Biagini P, Bertotto A (1995) High levels of the soluble form of CD30 molecule in rheumatoid arthritis (RA) are expression of CD30+ T cell involvement in the inflamed joints. Clin Exp Immunol 102:547 550

Germann T, Hess H, Szeliga J, Rude E (1996) Characterization of the adjuvant effect of IL-12 and efficacy of IL-12 inhibitors in type II collagen-induced arthritis. Ann NY Acad Sci 795:227 240

Germann T, Rude E, Mattner F, Gately MK (1995) The IL-12 p40 homodimer as a specific antagonist of the IL-12 heterodimer. Immunol Today 16:500 501

Gillessen S, Carvajal D, Ling P, Podlaski FJ, Stremlo DL, Familletti PC, Gubler U, Presky DH, Stern AS, Gately MK (1995) Mouse interleukin-12 (IL-12) p40 homodimer: a potent IL-12 antagonist. Eur J Immunol 25:200 206

Gorham JD, Güler ML, Steen RG, Mackey AJ, Daly MJ, Frederick K, Dietrich WF, Murphy KM (1996) Genetic mapping of a murine locus controlling development of T helper 1/T helper 2 type responses. Proc Natl Acad Sci USA 93:12467 12472

Grewal IS, Grewal KD, Wong FS, Picarella DE, Janeway CA Jr, Flavell RA (1996) Local expression of transgene encoded TNF alpha in islets prevents autoimmune diabetes in nonobese diabetic (NOD) mice by preventing the development of auto-reactive islet-specific T cells. J Exp Med 184:1963 1974

Guerder S, Picarella DE, Linsley PS, Flavell RA (1994) Costimulator B7-1 confers antigen-presenting-cell function to parenchymal tissue and in conjunction with tumor necrosis factor alpha leads to auto-immunity in transgenic mice. Proc Natl Acad Sci USA 91:5138 5142

Hagopian WA, Karlsen AE, Gottsater A, Landin-Olsson M, Grubin CE, Sundkvist G, Petersen JS, Boel E, Dyrberg T, Lernmark A (1993) Quantitative assay using recombinant human islet glutamic acid decarboxylase (GAD65) shows that 64 K autoantibody positivity at onset predicts diabetes type. J Clin Invest 91:368 374

Han HS, Jun HS, Utsugi T, Yoon JW (1996) A new type of CD4 + suppressor T cell completely prevents spontaneous autoimmune diabetes and recurrent diabetes in syngeneic islet-transplanted NOD mice. J Autoimmun 9:331 339

Hancock WW, Polanski M, Zhang J, Blogg N, Weiner HL (1995) Suppression of insulitis in nonobese diabetic (NOD) mice by oral insulin administration is associated with selective expression of interleukin-4 and -10, transforming growth factor-beta, and prostaglandin-E. Amer J Pathol 147:1193 1199

Hanson MS, Cetkovic-Cvrlje M, Ramiya VK, Atkinson MA, Maclaren NK, Singh B, Elliott JF, Serreze DV, Leiter EH (1996) Quantitative thresholds of MHC class II I-E expressed on hematopoietically derived antigen-presenting cells in transgenic NOD/Lt mice determine level of diabetes resistance and indicate mechanism of protection. J Immunol 157:1279 1287

Harding FA, McArthur JG, Gross JA, Raulet DH, Allison JP (1992) CD28-mediated signalling co-stimulates murine T cells and prevents induction of anergy in T-cell clones. Nature 356:607 609

Haskins K, Wegmann D (1996) Diabetogenic T-cell clones. Diabetes 45:1299 1305

Healey D, Ozegbe P, Arden S, Chandler P, Hutton J, Cooke A (1995) In vivo activity and in vitro specificity of CD4 + Th1 and Th2 cells derived from the spleens of diabetic NOD mice. J Clin Invest 95:2979 2985

Heremans H, Dillen C, Groenen M, Martens E, Billiau A (1996) Chronic relapsing experimental auto-immune encephalomyelitis (CREAE) in mice: enhancement by monoclonal antibodies against interferon-gamma. Eur J Immunol 26:2393 2398

Herold KC, Vezys V, Sun Q, Viktora D, Seung E, Reiner S, Brown DR (1996) Regulation of cytokine production during development of autoimmune diabetes induced with multiple low doses of streptozotocin. J Immunol 156:3521 3527

Herrera PL, Harlan DM, Fossati L, Izui S, Huarte J, Orci L, Vassalli JD, Vassalli P (1994) A CD8 + T-lymphocyte-mediated and CD4 + T-lymphocyte-independent autoimmune diabetes of early onset in transgenic mice. Diabetologia 37:1277 1279

Hess H, Gately MK, Rude E, Schmitt E, Szeliga J, Germann T (1996) High doses of interleukin-12 inhibit the development of joint disease in DBA/1 mice immunized with type II collagen in complete Freund's adjuvant. Eur J Immunol 26:187 191

Hesse M, Bayrak S, Mitchison A (1996) Protective major histocompatibility complex genes and the role of interleukin-4 in collagen-induced arthritis. Eur J Immunol 26:3234 3237

Hosken NA, Shibuya K, Heath AW, Murphy KM, O'Garra A (1995) The effect of antigen dose on CD4 + T helper cell phenotype development in a T cell receptor-αβ-transgenic model. J Exp Med 182:1579 1584

Howell M, Winters S, Olee T, Powell H, Carlo D, Brostoff S (1989) Vaccination against experimental allergic encephalomyelitis with T cell receptor peptides. Science 246:668

Hultgren B, Huang X, Dybdal N, Stewart TA (1996) Genetic absence of gamma-interferon delays but does not prevent diabetes in NOD mice. Diabetes 45:812 817

Hunger RE, Carnaud C, Garcia I, Vassalli P, Mueller C (1997) Prevention of autoimmune diabetes mellitus in NOD mice by transgenic expression of soluble tumor necrosis factor receptor p55. Eur J Immunol 27:255 261

Inobe JI, Chen Y, Weiner HL (1996) In vivo administration of IL-4 induces TGF-beta-producing cells and protects animals from experimental autoimmune encephalomyelitis. Ann NY Acad Sci 778:390 392

Interferon-β multiple sclerosis study group (1993) Interferon beta-1b is effective in relapsing-remitting multiple sclerosis. I. Clinical results of a multicenter, randomized, double-blind, placebo-controlled trial. Neurology 43:665

Jacob CO, Aiso S, Michie SA, McDevitt HO, Acha-Orbea H (1990) Prevention of diabetes in nonobese diabetic mice by tumor necrosis factor (TNF): similarities between TNF-alpha and interleukin 1. Proc Natl Acad Sci USA 87:968 972

Jacob CO, Holoshitz J, Van der Meide P, Strober S, McDevitt HO (1989) Heterogeneous effects of IFN-gamma in adjuvant arthritis. J Immunol 142:1500 1505

Johns L, Sriram S (1993) Experimental allergic encephalomyelitis: Neutralizing antibody to TGFβ1 enhances the clinical severity of the disease. J Neuroimmunol 47:1

Joosten LA, Helsen MM, van de Loo FA, van den Berg WB (1996) Anticytokine treatment of established type II collagen-induced arthritis in DBA/1 mice. A comparative study using anti-TNF alpha, anti-IL-1 alpha/beta, and IL-1Ra. Arthr Rheum 39:797 809

Kallmann BA, Huther M, Tubes M, Feldkamp J, Bertrams J, Gries FA, Lampeter EF, Kolb H (1997) Systemic bias of cytokine production toward cell-mediated immune regulation in IDDM and toward humoral immunity in Graves' disease. Diabetes 46:237 243

Karin N, Mitchell DJ, Brocke S, Ling N, Steinman L (1994) Reversal of experimental autoimmune encephalomyelitis by a soluble peptide variant of a myelin basic protein epitope: T cell receptor antagonism and reduction of interferon gamma and tumor necrosis factor alpha production. J Exp Med 180:2227 2237

Karpus WJ, Gould KE, Swanborg RH (1992) CD4+ suppressor cells of autoimmune encephalomyelitis respond to T cell receptor-associated determinants on effector cells by interleukin-4 secretion. Eur J Immunol 22:1757 1763

Karpus WJ, Peterson JD, Miller SD (1994) Anergy in vivo: down-regulation of antigen-specific CD4+ Th1 but not Th2 cytokine responses. Int Immunol 6:721 730

Katz JD, Benoist C, Mathis D (1995) T helper cell subsets in insulin-dependent diabetes. Science 268:1185 1188

Kaufman DL, Clare-Salzler M, Tian J, Forsthuber T, Ting GS, Robinson P, Atkinson MA, Sercarz EE, Tobin AJ, Lehmann PV (1993) Spontaneous loss of T-cell tolerance to glutamic acid decarboxylase in murine insulin-dependent diabetes. Nature 366:69 72

Kennedy MK, Torrance DS, Picha KS, Mohler KM (1992) Analysis of cytokine mRNA expression in the central nervous system of mice with experimental autoimmune encephalomyelitis reveals that IL-10 mRNA expression correlates with recovery. J Immunol 149:2496 2505

Khoruts A, Miller SD, Jenkins MK (1995) Neuroantigen-specific Th2 cells are inefficient suppressors of experimental autoimmune encephalomyelitis induced by effector Th1 cells. J Immunol 155:5011 5017

Khoury SJ, Hancock WW, Weiner HL (1992) Oral tolerance to myelin basic protein and natural recovery from experimental autoimmune encephalomyelitis are associated with downregulation of inflammatory cytokines and differential upregulation of transforming growth factor beta, interleukin 4, and prostaglandin E expression in the brain. J Exp Med 176(5):1355 1364

Kotzin BL, Karuturi S, Chou YK, Lafferty J, Forrester JM, Better M, Nedwin GE, Offner H, Vandenbark AA (1991) Preferential T-cell receptor beta-chain variable gene use in myelin basic protein-reactive T-cell clones from patients with multiple sclerosis. Proc Natl Acad Sci USA 88:9161 9165

Krakowski M, Owens T (1996) Interferon-gamma confers resistance to experimental allergic encephalomyelitis. Eur J Immunol 26:1641 1646

Krummel MF, Allison JP (1996) CTLA-4 engagement inhibits IL-2 accumulation and cell cycle progression upon activation of resting T cells. J Exp Med 183:2533 2540

Kuchroo VK, Das MP, Brown JA, Ranger AM, Zamvil SS, Sobel RA, Weiner HL, Nabavi N, Glimcher LH (1995) B7-1 and B7-2 costimulatory molecules activate differentially the Th1/Th2 developmental pathways: application to autoimmune disease therapy. Cell 80:707 18

Kuchroo VK, Greer JM, Kaul D, Ishioka G, Franco A, Sette A, Sobel RA, Lees MB (1994) A single TCR antagonist peptide inhibits experimental allergic encephalomyelitis mediated by a diverse T cell repertoire. J Immunol 153:3326 3336

Kuchroo VK, Martin CA, Greer JM, Ju ST, Sobel RA, Dorf ME (1993) Cytokines and adhesion molecules contribute to the ability of myelin proteolipid protein-specific T cell clones to mediate experimental allergic encephalomyelitis. J Immunol 151:4371 4382

Kuchroo VK, Sobel RA, Laning JC, Martin CA, Greenfield E, Dorf ME, Lees MB (1992) Experimental allergic encephalomyelitis mediated by cloned T cells specific for a synthetic peptide of myelin proteolipid protein. Fine specificity and T cell receptor V beta usage. J Immunol 148:3776 3782

Kumar V, Tabibiazar R, Geysen HM, Sercarz E (1995) Immunodominant framework region 3 peptide from TCR V beta 8.2 chain controls murine experimental autoimmune encephalomyelitis. J Immunol 154:1941 1950

Lamont AG, Adorini L (1996) IL-12: a key cytokine in immune regulation. Immunol Today 17:214 217

Lee MS, von Herrath M, Reiser H, Oldstone MB, Sarvetnick N (1995) Sensitization to self (virus) antigen by in situ expression of murine interferon-gamma. J Clin Invest 95:486 492

Lenschow DJ, Herold KC, Rhee L, Patel B, Koons A, Qin HY, Fuchs E, Singh B, Thompson CB, Bluestone JA (1996) CD28/B7 regulation of Th1 and Th2 subsets in the development of autoimmune diabetes. Immunity 5:285 293

Lenschow DJ, Ho SC, Sattar H, Rhee L, Gray G, Nabavi N, Herold KC, Bluestone JA (1995) Differential effects of anti-B7-1 and anti-B7-2 monoclonal antibody treatment on the development of diabetes in the nonobese diabetic mouse. J Exp Med 181:1145 1155

Leonard JP, Waldburger KE, Goldman SJ (1995) Prevention of experimental autoimmune encephalomyelitis by antibodies against interleukin 12. J Exp Med 181:381 386

Leonard JP, Waldburger KE, Goldman SJ (1996) Regulation of experimental autoimmune encephalomyelitis by interleukin-12. Ann NY Acad Sci 795:216 226

Liblau RS, Pearson CI, Shokat K, Tisch R, Yang XD, McDevitt HO (1994) High-dose soluble antigen: peripheral T-cell proliferation or apoptosis. Immunol Rev 142:193 208

Liblau RS, Singer SM, McDevitt HO (1995) Th1 and Th2 CD4 + T cells in the pathogenesis of organ-specific autoimmune diseases. Immunol Today 16:34 38

Lider O, Reshef T, Beraud E, Ben-Nun A, Cohen I (1988) Anti-idiotypic network induced by T cell vaccination against experimental autoimmune encephalomyelitis. Science 239:181

Liu GY, Wraith DC (1995) Affinity for class II MHC determines the extent to which soluble peptides tolerize autoreactive T cells in naive and primed adult mice implications for autoimmunity. Int Immunol 7:1255 1263

Lund T, O'Reilly L, Hutchings P, Kanagawa O, Simpson E, Gravely R, Chandler P, Dyson J, Picard JK, Edwards A, et al. (1990) Prevention of insulin-dependent diabetes mellitus in nonobese diabetic mice by transgenes encoding modified I-A beta-chain or normal I-E alpha-chain. Nature 345:727 729

Lyons DS, Lieberman SA, Hampl J, Boniface JJ, Chien Y, Berg LJ, Davis MM (1996) A TCR binds to antagonist ligands with lower affinities and faster dissociation rates than to agonists. Immunity 5:53 61

Magram J, Sfarra J, Connaughton S, Faherty D, Warrier R, Carvajal D, Wu CY, Stewart C, Sarmiento U, Gately MK (1996) IL-12-deficient mice are defective but not devoid of type 1 cytokine responses. Ann NY Acad Sci 795:60 70

Martin D, Near SL, Bendele A, Russell DA (1995) Inhibition of tumor necrosis factor is protective against neurologic dysfunction after active immunization of Lewis rats with myelin basic protein. Exp Neurol 131:221 228

Mauri C, Williams RO, Walmsley M, Feldmann M (1996) Relationship between Th1/Th2 cytokine patterns and the arthritogenic response in collagen-induced arthritis. Eur J Immunol 26:1511 1518

McIntyre KW, Shuster DJ, Gillooly KM, Warrier RR, Connaughton SE, Hall LB, Arp LH, Gately MK, Magram J (1996) Reduced incidence and severity of collagen-induced arthritis in interleukin-12-deficient mice. Eur J Immunol 26:2933 2938

Metzler B, Wraith DC (1993) Inhibition of experimental autoimmune encephalomyelitis by inhalation but not oral administration of the encephalitogenic peptide: influence of MHC binding affinity. Int Immunol 5:1159 1165

Miller SD, Vanderlugt CL, Lenschow DJ, Pope JG, Karandikar NJ, Dal Canto MC, Bluestone JA (1995) Blockade of CD28/B7-1 interaction prevents epitope spreading and clinical relapses of murine EAE. Immunity 3:739 745

Miller JFAP, Morahan G (1992) Peripheral T cell tolerance. Ann Rev Immunol 10:51

Miltenberg AM, van Laar JM, de Kuiper R, Daha MR, Breedveld FC (1992) T cells cloned from human rheumatoid synovial membrane functionally represent the Th1 subset. Scand J Immunol 35:603 610

Mordes JP, Schirf B, Roipko D, Greiner DL, Weiner H, Nelson P, Rossini AA (1996) Oral insulin does not prevent insulin-dependent diabetes mellitus in BB rats. Ann NY Acad Sci 778:418 421

Morgan DJ, Liblau R, Scott B, Fleck S, McDevitt HO, Sarvetnick N, Lo D, Sherman LA (1996) CD8(+) T cell-mediated spontaneous diabetes in neonatal mice. J Immunol 157:978 983

Mori L, Iselin S, De Libero G, Lesslauer W (1996) Attenuation of collagen-induced arthritis in 55-kDa TNF receptor type 1 (TNFR1)-IgG1-treated and TNFR1-deficient mice. J Immunol 157:3178 3182

Moritani M, Yoshimoto K, Ii S, Kondo M, Iwahana H, Yamaoka T, Sano T, Nakano N, Kikutani H, Itakura M (1996) Prevention of adoptively transferred diabetes in nonobese diabetic mice with IL-10-transduced islet-specific Th1 lymphocytes. A gene therapy model for autoimmune diabetes. J Clin Invest 98:1851 1859

Moritani M, Yoshimoto K, Tashiro F, Hashimoto C, Miyazaki J, Ii S, Kudo E, Iwahana H, Hayashi Y, Sano T, et al. (1994) Transgenic expression of IL-10 in pancreatic islet A cells accelerates autoimmune insulitis and diabetes in nonobese diabetic mice. Int Immunol 6:1927 1936

Mueller R, Lee MS, Sawyer SP, Sarvetnick N (1996) Transgenic expression of interleukin 10 in the pancreas renders resistant mice susceptible to low dose streptozotocin-induced diabetes. J Auto-immun 1996 9:151 158

Mueller D, Jenkins M, Schwartz R (1989) Clonal expansion versus functional clonal inactivation: A costimulatory signalling pathway determines the outcome of T cells antigen receptor occupancy. Ann Rev Immuno 17:445

Muir A, Peck A, Clare-Salzler M, Song YH, Cornelius J, Luchetta R, Krischer J, Maclaren N (1995) Insulin immunization of nonobese diabetic mice induces a protective insulitis characterized by di-minished intraislet interferon-gamma transcription. J Clin Invest 95:628 634

Myers LK, Stuart JM, Seyer JM, Kang AH (1989) Identification of an immunosuppressive epitope of type II collagen that confers protection against collagen-induced arthritis. J Exp Med 170:1999 2010

Nicholson LB, Greer JM, Sobel RA, Lees MB, Kuchroo VK (1995) An altered peptide ligand mediates immune deviation and prevents autoimmune encephalomyelitis. Immunity 3:397 405

Nicoletti F, Zaccone P, Di Marco R, Lunetta M, Magro G, Grasso S, Meroni P, Garotta G (1997) Prevention of spontaneous autoimmune diabetes in diabetes-prone BB rats by prophylactic treatment with antirat interferon-gamma antibody. Endocrinology 138:281 288

Nishimoto H, Kikutani H, Yamamura K, Kishimoto T (1987) Prevention of autoimmune insulitis by expression of I-E molecules in NOD mice. Nature 328:432 434

Noronha A, Toscas A, Jensen MA (1992) Contrasting effects of alpha, beta, and gamma interferons on nonspecific suppressor function in multiple sclerosis. Ann Neurol 31:103 106

Nossal GJ (1983) Cellular mechanisms of immunologic tolerance. Ann Rev Immunol 1:33 62

O'Garra A, Murphy K (1996) Role of cytokines in development of Th1 and Th2 cells. Chem Immunol 63:1 13

O'Garra A, Steinman L, Gijbels, K (1997) CD4 + T-cell subsets in auto-immunity Curr op Immunol 9:872 883

O'Hara RM Jr, Henderson SL, Nagelin A (1996) Prevention of a Th1 disease by a Th1 cytokine: IL-12 and diabetes in NOD mice. Ann NY Acad Sci 795:241 249

Offner H, Hashim GA, Vandenbark AA (1991) T cell receptor peptide therapy triggers autoregulation of experimental encephalomyelitis. Science 259:91

Ogata LC, Reilly C, Marconi LAM, Lo D (1995) Anti-IL-12 antibody can prevent autoimmune diabetes in an adoptive transfer murine model. J Cell Biochem Suppl 21A:150

Oksenberg JR, Panzara MA, Begovich AB, Mitchell D, Erlich HA, Murray RS, Shimonkevitz R, Sherritt M, Rothbard J, Bernard CCA, Steinman L (1993) Selection for T-cell receptor Vβ-Dβ-Jβ gene rearrangements with specificity for a myelin basic protein peptide in brain lesions of multiple sclerosis. Nature 362:68 70

Paganin C, Gerosa F, Peritt D, Paiola F, Scupoli MT, Aste-Amezaga M, Frank I, Trinchieri G (1996) Effect of interleukin-12 on the cytokine profile of human CD4 and CD8 T-cell clones. Ann NY Acad Sci 795:382 383

Panitch HS, Hirsch RL, Haley AS, Johnson KP (1987) Exacerbations of multiple sclerosis in patients treated with gamma interferon. Lancet 1:893 895

Panitch HS, Hooper CJ, Johnson KP (1980) CSF antibody to myelin basic protein. Measurement in patients with multiple sclerosis and subacute sclerosing panencephalitis. Arch Neurol 37:206 209

Paul WE, Seder RA (1994) Lymphocyte responses and cytokines. Cell 76:241 251

Pearson CI, van Ewijk W, McDevitt HO (1997) Induction of apoptosis and Th2 responses correlates with peptide affinity for the major histocompatibility complex in self-reactive T cell receptor transgenic mice. J Exp Med 185:583 599

Pennline KJ, Roque-Gaffney E, Monahan M (1994) Recombinant human IL-10 prevents the onset of diabetes in the nonobese diabetic mouse. Clin Immunol Immunopathol 71:169 175

Persson S, Mikulowska A, Narula S, O'Garra A, Holmdahl R (1996) Interleukin-10 suppresses the development of collagen type II-induced arthritis and ameliorates sustained arthritis in rats. Scand J Immunol 44:607 614

Petersen JS, Karlsen AE, Markholst H, Worsaae A, Dyrberg T, Michelsen B (1994) Neonatal tolerization with glutamic acid decarboxylase but not with bovine serum albumin delays the onset of diabetes in NOD mice. Diabetes 43:1478 1484

Peterson JD, Haskins K (1996) Transfer of diabetes in the NOD-scid mouse by CD4 T-cell clones. Differential requirement for CD8 T-cells. Diabetes 45:328 336

Peterson JD, Karpus WJ, Clatch RJ, Miller SD (1993) Split tolerance of Th1 and Th2 cells in tolerance to Theiler's murine encephalomyelitis virus. Eur J Immunol 23:46 55

Pfeiffer C, Stein J, Southwood S, Ketelaar H, Sette A, Bottomly K (1995) Altered peptide ligands can control CD4 T lymphocyte differentiation in vivo. J Exp Med 181:1569 1574

Piccotti JR, Chan SY, Li K, Eichwald EJ, Bishop DK (1997) Differential effects of IL-12 receptor blockade with IL-12 p40 homodimer on the induction of CD4 + and CD8 + IFN-gamma-producing cells. J Immunol 158:643 648

Pilstrom B, Bjork L, Bohme J (1995) Demonstration of a TH1 cytokine profile in the late phase of NOD insulitis. Cytokine 7:806 814

Powell MB, Mitchell D, Lederman J, Buckmeier J, Zamvil SS, Graham M, Ruddle NH, Steinman L (1990) Lymphotoxin and tumor necrosis factor-alpha production by myelin basic protein-specific T cell clones correlates with encephalitogenicity. Int Immunol 2:539 544

Qin HY, Sadelain MW, Hitchon C, Lauzon J, Singh B (1993) Complete Freund's adjuvant-induced T cells prevent the development and adoptive transfer of diabetes in nonobese diabetic mice. J Immunol 150:2072 2080

Quayle AJ, Chomarat P, Miossec P, Kjeldsen-Kragh J, Forre O, Natvig JB (1993) Rheumatoid inflammatory T-cell clones express mostly Th1 but also Th2 and mixed (Th0-like) cytokine patterns. Scand J Immunol 38:75 82

Rabinovitch A, Suarez-Pinzon WL, Lapchak PH, Meager A, Power RF (1995) Tumor necrosis factor mediates the protective effect of Freund's adjuvant against autoimmune diabetes in BB rats. J Autoimmun 8:357 366

Rabinovitch A, Suarez-Pinzon WL, Sorensen O, Bleackley RC, Power RF, Rajotte RV (1995) Combined therapy with interleukin-4 and interleukin-10 inhibit autoimmune diabetes recurrence in syngeneic islet-transplanted nonobese diabetic mice. Analysis of cytokine mRNA expression in the graft. Transplantation 60:368 374

Racke MK, Bonomo A, Scott DE, Cannella B, Levine A, Raine CS, Shevach EM, Rocken M (1994) Cytokine-induced immune deviation as a therapy for inflammatory autoimmune disease. J Exp Med 180:1961 1966

Racke MK, Sriram S, Carlino J, Cannella B, Raine CS, McFarlin DE (1993) Long-term treatment of chronic relapsing experimental allergic encephalomyelitis by transforming growth factor-β2. J Neuroimmunol 46:175

Raine CS (1995) Multiple sclerosis: TNF revisited, with promise. Nature Med 1:211 214

Ramanathan S, de Kozak Y, Saoudi A, Goureau O, Van der Meide PH, Druet P, Bellon B (1996) Recombinant IL-4 aggravates experimental autoimmune uveoretinitis in rats. J Immunol 157:2209 2215

Ramiya VK, Shang XZ, Pharis PG, Wasserfall CH, Stabler TV, Muir AB, Schatz DA, Maclaren NK (1996) Antigen based therapies to prevent diabetes in NOD mice. J Autoimmun 9:349 356

Ramsdell F, Seaman MS, Miller RE, Picha KS, Kennedy MK, Lynch DH (1994) Differential ability of Th1 and Th2 T cells to express Fas ligand and to undergo activation-induced cell death. Int Immunol 6:1545 553

Ranger AM, Das MP, Kuchroo VK, Glimcher LH (1996) B7-2 (CD86) is essential for the development of IL-4-producing T cells. Int Immunol 8:1549 1560

Rankin ECC, Choy EHS, Kassimos D, Kingsley GH, Sopwith SM, Isenberg D, Panayi GS (1995) The therapeutic effects of an engineered human anti-tumour necrosis factor alpha antibodies (CDP571) in rheumatoid arthritis. Br J Rheumatol 34:334 342

Rapoport MJ, Jaramillo A, Zipris D, Lazarus AH, Serreze DV, Leiter EH, Cyopick P, Danska JS, Delovitch TL (1993) Interleukin 4 reverses T cell proliferative unresponsiveness and prevents the onset of diabetes in nonobese diabetic mice. J Exp Med 178:87–99

Rep MH, Hintzen RQ, Polman CH, van Lier RA (1996) Recombinant interferon-beta blocks proliferation but enhances interleukin-10 secretion by activated human T-cells. J Neuroimmunol 67:111–118

Ridge JP, Fuchs EJ, Matzinger P (1996) Neonatal tolerance revisited: turning on newborn T cells with dendritic cells. Science 271:1723–1726

Rott O, Fleischer B, Cash E (1994) Interleukin-10 prevents experimental allergic encephalomyelitis in rats. Eur J Immunol 24:1434–440

Ruddle NH, Bergman CM, McGrath KM, Lingenheld EG, Grunnet ML, Padula SJ, Clark RB (1990) An antibody to lymphotoxin and tumor necrosis factor prevents transfer of experimental allergic encephalomyelitis. J Exp Med 172:1193–2000

Rulifson IC, Sperling AI, Fields PE, Fitch FW, Bluestone JA (1997) CD28 costimulation promotes the production of Th2 cytokines. J Immunol 158:658–665

Sa P, Rivereau AS, Granier C, Haertlé T, Martignat L (1996) Immunization of nonobese diabetic (NOD) mice with glutamic acid decarboxylase-derived peptide 524–543 reduces cyclophosphamide-accelerated diabetes. Clin Exp Immunol 105:330–337

Sakai K, Sinha AA, Mitchell DJ, Zamvil SS, Rothbard JB, McDevitt HO, Steinman L (1988) Involvement of distinct murine T-cell receptors in the autoimmune encephalitogenic response to nested epitopes of myelin basic protein. Proc Natl Acad Sci USA 85:8608–8612

Samson MF, Smilek DE (1995) Reversal of acute experimental autoimmune encephalomyelitis and prevention of relapses by treatment with a myelin basic protein peptide analogue modified to form long-lived peptide-MHC complexes. J Immunol 155:2737–2746

Santambrogio L, Crisi GM, Leu J, Hochwald GM, Ryan T, Thorbecke GJ (1995) Tolerogenic forms of auto-antigens and cytokines in the induction of resistance to experimental allergic encephalomyelitis. J Neuroimmunol 58:211–222

Santambrogio L, Hochwald GM, Leu CH, Thorbecke GJ (1993a) Antagonistic effects of endogenous and exogenous TGF-beta and TNF on auto-immune diseases in mice. Immunopharm Immunotoxicol 15:461–478

Santambrogio L, Hochwald GM, Saxena B, Leu CH, Martz JE, Carlino JA, Ruddle NH, Palladino MA, Gold LI, Thorbecke GJ (1993b) Studies on the mechanisms by which transforming growth factor-beta (TGF-beta) protects against allergic encephalomyelitis. Antagonism between TGF-beta and tumor necrosis factor. J Immunol 151:1116–1127

Sarvetnick N, Shizuru J, Liggitt D, Martin L, McIntyre B, Gregory A, Parslow T, Stewart T (1990) Loss of pancreatic islet tolerance induced by beta-cell expression of interferon-gamma. Nature 346:844–847

Sarzotti M, Robbins DS, Hoffman PM (1996) Induction of protective CTL responses in newborn mice by a murine retrovirus. Science 271:1726–1728

Satoh J, Seino H, Shintani S, Tanaka S, Ohteki T, Masuda T, Nobunaga T, Toyota T (1990) Inhibition of type 1 diabetes in BB rats with recombinant human tumor necrosis factor-alpha. J Immunol 145:1395–1399

Schulze-Koops H, Lipsky PE, Kavanaugh AF, Davis LS (1995) Elevated Th1- or Th0-like cytokine mRNA in peripheral circulation of patients with rheumatoid arthritis. Modulation by treatment with anti-ICAM-1 correlates with clinical benefit. J Immunol 155:5029–5037

Scolding NJ, Morgan BP, Houston WA, Linington C, Campbell AK, Compston DA (1989) Vesicular removal by oligodendrocytes of membrane attack complexes formed by activated complement. Nature 339:620–622

Scott B, Liblau R, Degermann S, Marconi LA, Ogata L, Caton AJ, McDevitt HO, Lo D (1994) A role for non-MHC genetic polymorphism in susceptibility to spontaneous autoimmunity. Immunity 1:73–83

Seino H, Takahashi K, Satoh J, Zhu XP, Sagara M, Masuda T, Nobunaga T, Funahashi I, Kajikawa T, Toyota T (1993) Prevention of autoimmune diabetes with lymphotoxin in NOD mice. Diabetes 42:398–404

Selmaj K, Papierz W, Glabinski A, Kohno T (1995) Prevention of chronic relapsing experimental autoimmune encephalomyelitis by soluble tumor necrosis factor receptor I. J Neuroimmunol 56:135–141

Sher A, Coffman RL (1992) Regulation of immunity to parasites by T cells and T cell-derived cytokines. Ann Rev Immunol 10:385–409

Shimada A, Charlton B, Taylor-Edwards C, Fathman CG (1996) Beta-cell destruction may be a late consequence of the autoimmune process in nonobese diabetic mice. Diabetes. 45:1063–1067

Singer SM, Tisch R, Yang XD, McDevitt HO (1993) An Aβd transgene prevents diabetes in nonobese diabetic mice by inducing regulatory T cells. Proc Natl Acad Sci USA 90:9566–570

Singer SM, Umetsu DT, McDevitt HO (1996) High copy number I-Aβ transgenes induce production of IgE through an interleukin 4-dependent mechanism. Proc Natl Acad Sci USA 93:2947 2952

Singh RR, Hahn BH, Sercarz EE (1996) Neonatal peptide exposure can prime T cells and, upon subsequent immunization, induce their immune deviation: implications for antibody vs. T cell-mediated autoimmunity. J Exp Med 183:1613 1621

Slattery RM, Kjer-Nielsen L, Allison J, Charlton B, Mandel TE, Miller JF (1990) Prevention of diabetes in nonobese diabetic I-Ak transgenic mice. Nature 345:724 726

Sloan-Lancaster J, Allen PM (1996) Altered peptide ligand-induced partial T cell activation: molecular mechanisms and role in T cell biology. Ann Rev Immunol 14:1 27

Smilek DE, Wraith DC, Hodgkinson S, Dwivedy S, Steinman L, McDevitt HO (1991) A single amino acid change in a myelin basic protein peptide confers the capacity to prevent rather than induce experimental autoimmune encephalomyelitis. Proc Natl Acad Sci USA. 88:9633 9637

Steinman L (1996) Multiple sclerosis: A coordinated immunological attack against myelin in the central nervous system. Cell 85:299 302

Stephens LA, Kay TW (1995) Pancreatic expression of B7 co-stimulatory molecules in the nonobese diabetic mouse. Int Immunol 7:1885 1895

Stumbles P, Mason D (1995) Activation of CD4 + T cells in the presence of a nondepleting monoclonal antibody to CD4 induces a Th2-type response in vitro. J Exp Med 182:5 13

Szeliga J, Hess H, Rude E, Schmitt E, Germann T (1996) IL-12 promotes cellular but not humoral type II collagen-specific Th1-type responses in C57BL/6 and B10.Q mice and fails to induce arthritis. Int Immunol 8:1221 1227

Takahashi K, Satoh J, Seino H, Zhu XP, Sagara M, Masuda T, Toyota T (1993) Prevention of type I diabetes with lymphotoxin in BB rats. Clin Immunol Immunopathol 69:318 332

Tanaka Y, Otsuka T, Hotokebuchi T, Miyahara H, Nakashima H, Kuga S, Nemoto Y, Niiro H, Niho Y (1996) Effect of IL-10 on collagen-induced arthritis in mice. Inflammation Res 45:283 288

Tian J, Atkinson MA, Clare-Salzler M, Herschenfeld A, Forsthuber T, Lehmann PV, Kaufman DL (1996a) Nasal administration of glutamate decarboxylase (GAD65) peptides induces Th2 responses and prevents murine insulin-dependent diabetes. J Exp Med 183:1561 1567

Tian J, Clare-Salzler M, Herschenfeld A, Middleton B, Newman D, Mueller R, Arita S, Evans C, Atkinson MA, Mullen Y, et al. (1996b) Modulating autoimmune responses to GAD inhibits disease progression and prolongs islet graft survival in diabetes-prone mice. Nature Med 2:1348 1353

Tisch R, Yang XD, Liblau RS, McDevitt HO (1994) Administering glutamic acid decarboxylase to NOD mice prevents diabetes. J Autoimmun 7:845 850

Tisch R, Yang XD, Singer SM, Liblau RS, Fugger L, McDevitt HO (1993) Immune response to glutamic acid decarboxylase correlates with insulitis in nonobese diabetic mice. Nature 366:72 75

Tisch R, McDevitt HO (1996) Insulin-dependent diabetes mellitus. Cell 85:291 297

Tisch R, Liblau RS, Yang X-D, Liblau P, McDevitt HO (1998) Induction of GAD65-specific regulatory T-cells inhibits ongoing autoimmune diabetes in nonobese diabetic mice. Diabetes 47:894 899

Todd JA, Aitman TJ, Cornall RJ, Ghosh S, Hall JR, Hearne DM, Knight AM, Love JM, McAleer MA, Prins JB, Rodrigues N, Lathrop M, Pressey A, DeLarato NH, Peterson LB, Wicker LS (1991) Genetic analysis of autoimmune type 1 diabetes mellitus in mice. Nature 351:542 547

Trembleau S, Penna G, Bosi E, Mortara A, Gately MK, Adorini L (1995) Interleukin 12 administration induces T helper type 1 cells and accelerates autoimmune diabetes in NOD mice. J Exp Med 181:817 821

Urban J, Kumar V, Kono D, Gomez C, Horvath S, Clayton J, Ando D, Sercarz E, Hood L (1988) Restricted use of T cell receptor V genes in murine autoimmune encephalomyelitis raises possibilities for antibody therapy. Cell 54:577 592

van der Veen RC, Kapp JA, Trotter JL (1993) Fine-specificity differences in the recognition of an encephalitogenic peptide by T helper 1 and 2 cells. J Neuroimmunol 48:221 226

van Oosten BW, Barkhof F, Truyen L, Boringa JB, Bertelsmann FW, von Blomberg BM, Woody JN, Hartung HP, Polman CH (1996) Increased MRI activity and immune activation in two multiple sclerosis patients treated with the monoclonal anti-tumor necrosis factor antibody cA2. Neurology 47:1531 1534

van Roon JA, van Roy JL, Duits A, Lafeber FP, Bijlsma JW (1995) Proinflammatory cytokine production and cartilage damage due to rheumatoid synovial T helper-1 activation is inhibited by interleukin-4. Ann Rheum Dis 54:836 840

Vandenbark A, Hashin G, Offner H (1989) Immunization with a synthetic T-cell receptor V-region peptide protects against experimental autoimmune encephalomyelitis. Nature 341:541

von Herrath MG, Oldstone MBA (1997) Interferon-γ is essential for destruction of β cells and development of insulin-dependent diabetes mellitus. J Exp Med 185:460 465

von Herrath MG, Guerder S, Lewicki H, Flavell RA, Oldstone MBA (1995) Coexpression of B7-1 and viral ("self") transgenes and pancreatic β cells can break peripheral ignorance and lead to spontaneous autoimmune diabetes. Immunity 3:727 738

Voorthuis JAC, Uitdenhaag BMJ, de Groot CJA, Goede HP, van der Meide PH, Dijkstra CD (1990) Suppression of experimental allergic encephalomyelitis by intraventricular administration of interferon-gamma in Lewis rats. Clin Exp Immunol 81:183

Walmsley M, Katsikis PD, Abney E, Parry S, Williams RO, Maini RN, Feldmann M (1996) Interleukin-10 inhibition of the progression of established collagen-induced arthritis. Arth Rheum 39:495 503

Wegmann DR, Norbury-Glaser M, Daniel D (1994) Insulin-specific T cells are a predominant component of islet infiltrates in pre-diabetic NOD mice. Eur J Immunol 24:1853 1857

Weigle WO (1973) Immunological unresponsiveness. Adv Immunol 16:61

Weiner HL, Friedman A, Miller A, Khoury SJ, al-Sabbagh A, Santos L, Sayegh M, Nussenblatt RB, Trentham DE, Hafler DA (1994) Oral tolerance: immunologic mechanisms and treatment of animal and human organ-specific autoimmune diseases by oral administration of autoantigens. Ann Rev Immunol 12:809 837

Weiner HL, Mackin GA, Matsui M, Orav EJ, Khoury SJ, Dawson DM, Hafler DA (1993) Double-blind pilot trial of oral tolerization with myelin antigens in multiple sclerosis. Science 259:1321 1324

Wherrett DK, Singer SM, McDevitt HO (1997) Reduction in diabetes incidence in an I-Ag7 transgenic nonobese diabetic mouse line. Diabetes 46:1970 1974

Whitacre CC, Gienapp IE, Orosz CG, Bitar DM (1991) Oral tolerance in experimental autoimmune encephalomyelitis. III. Evidence for clonal anergy. J Immunol 147:2155 2163

Willenborg DO, Fordham SA, Cowden WB, Ramshaw IA (1995) Cytokines and murine autoimmune encephalomyelitis: inhibition or enhancement of disease with antibodies to select cytokines, or by delivery of exogenous cytokines using a recombinant vaccinia virus system. Scand J Immunol 41:31 41

Willenborg DO, Fordham S, Bernard CC, Cowden WB, Ramshaw IA (1996a) IFN-gamma plays a critical down-regulatory role in the induction and effector phase of myelin oligodendrocyte glycoprotein-induced autoimmune encephalomyelitis. J Immunol 157:3223 3227

Willenborg DO, Staykova MA, Miyasaka M (1996b) Short term treatment with soluble neuroantigen and anti-CD11a (LFA-1) protects rats against autoimmune encephalomyelitis: treatment abrogates autoimmune disease but not autoimmunity. J Immunol 157:1973 1980

Wogensen L, Lee MS, Sarvetnick N (1994) Production of interleukin 10 by islet cells accelerates immune-mediated destruction of β-cells in nonobese diabetic mice. J Exp Med 179:1379 1384

Wong S, Guerder S, Visintin I, Reich EP, Swenson KE, Flavell RA, Janeway CA (1995) Expression of the co-stimulator molecule B7-1 in pancreatic beta-cells accelerates diabetes in the NOD mouse. Diabetes 44:326 329

Wong FS, Visintin I, Wen L, Flavell RA, Janeway CA (1996) CD8 T cell clones from young nonobese diabetic (NOD) islets can transfer rapid onset of diabetes in NOD mice in the absence of CD4 cells. J Exp Med 183:67 76

Wucherpfennig KW, Ota K, Endo N, Seidman JG, Rosenzweig A, Weiner HL, Hafler DA (1990) Shared human T cell recpeotr V β usage to immunodominant regions of myelin basic protein. Science 1016 1019

Yang X-D, Tisch R, Singer SM, Cao ZA, Liblau RS, Schreiber RD, McDevitt HO (1994) Effect of tumor necrosis factor a on insulin-dependent diabetes mellitus in NOD mice. I. The early development of autoimmunity and the diabetogenic process. J Exp Med 180:995 1004

Zaller DM, Osman G, Kanagawa O, Hood L (1990) Prevention and treatment of murine experimental allergic encephalomyelitis with T cell receptor V beta-specific antibodies. J Exp Med 171:1943 1955

Zamvil SS, Steinman L (1990) The T lymphocyte in experimental allergic encephalomyelitis. Ann Rev Immunol 8:579 621

Zipris D, Greiner DL, Malkani S, Whalen B, Mordes JP, Rossini AA (1996) Cytokine gene expression in islets and thyroids of BB rats. IFN-gamma and IL-12p40 mRNA increase with age in both diabetic and insulin-treated nondiabetic BB rats. J Immunol 156:1315 1321

Manipulation of Th Responses by Oral Tolerance

G. Garcia and H.L. Weiner

1	Introduction	123
2	Antigen Uptake and Processing	124
3	Cytokine Milieu	125
4	Mechanisms of Oral Tolerance	126
5	Bystander Suppression and Oral Tolerance	129
6	Modulation of Oral Tolerance	131
7	Manipulation of Th Responses in Experimental Models of Autoimmunity	132
8	Experimental Autoimmune Encephalomyelitis	133
9	Insulin-Dependent Diabetes Mellitus	134
10	Collagen-Induced Arthritis	135
11	Treatment of Autoimmune Diseases in Humans	137
12	Future Directions	139
References		139

1 Introduction

The gastrointestinal tract is the major site of antigenic contact in the body. A complex and precise gastrointestinal immune system allows the accomplishment of two distinct features: absorption of nutrients avoiding hypersensitivity to food antigens and exclusion of microbiologic pathogens (KAGNOFF 1996; NEWBY 1984; STROBEL 1990).

Oral tolerance is a long recognized method of inducing immune tolerance. It refers to the observation that if one feeds a protein and then immunizes with the same protein, a state of systemic hyporesponsiveness to the fed protein exists (CHALLACOMBE and TOMASI 1980; CHASE 1946; MOWAT 1987; RICHMAN et al. 1978; VAZ et al. 1977; WEINER 1997). It is thought to be the mechanism that prevents intestinal hypersensitivity reactions to food antigens.

Center for Neurologic Diseases, Brigham and Women's Hospital, Harvard Medical School, 77 Avenue Louis Pasteur, HIM 730, Boston, MA 02215, USA

In recent years, as more has been learned about the general mechanisms of immune tolerance, investigators have begun to apply oral tolerance as a method to manipulate injurious immune responses, primarily in the area of autoimmune diseases, although its applications appear broader and have included transplantation as well. This area has gained intense interest and is likely to grow since manipulation of systemic immune responses via the mucosal immune system has major practical and theoretical advantages (WEINER and MAYER 1996). In this chapter, we will describe the organization of the gut-associated lymphoid tissue (GALT) and discuss how its unique properties influence the immunologic events that occur after administration of antigens by oral route.

2 Antigen Uptake and Processing

Uptake of macromolecules, particulate antigens and microorganisms across intestinal epithelia can occur only by active transepithelial vesicular transport, and this is restricted by multiple mechanisms including local secretions containing mucins and secretory IgA antibodies; rigid, closely packed microvilli; and the glycocalyx, which is a thick layer of membrane anchored proteins and several enzymes (NEUTRA and KRAEHENBUHL 1996).

The gut associated lymphoid tissue or GALT can be divided into layers (Fig. 1). The first layer consists of intraepithelial lymphocytes (IELs) which reside within the epithelium itself, above the basement membrane. Beneath this layer, in the lamina propria located between the epithelium and submucosa, reside the lamina propria lymphocytes (LPLs). Finally, an organized lymphoid tissue in the form of Peyer's patches (PP) is found predominantly in the small bowel and appendix (ABREU-MARTIN and TARGAN 1996).

The PP have a specialized epithelium, known as M cells, capable of transporting antigens across the mucosal barrier. In M cells, unlike most epithelial cells, transepithelial vesicular transport is the major pathway for endocytosed material. Most (but not all) M cells lack the uniform thick glycocalyx seen on enterocytes (NEUTRA and KRAEHENBUHL 1996).

B and T cells are present in PP along with macrophages. There is evidence that T cells can interact with antigen presenting cells in the PP and then migrate to mesenteric lymph nodes and reach the peripheral immune system as well as other mucosal tissues of the body (Fig. 1). PP are a major source of IgA-producing B cells, and T cells from PP have been reported to preferentially induce B cells into plasma cells secreting IgA (MACDONALD 1983). As we will discuss later, PP have also been shown to be the site where regulatory cells are generated which mediate the active suppression component of oral tolerance (MATTINGLY and WAKSMAN 1978; MILLER and HANSON 1979; NGAN and KIND 1978; RICHMAN et al. 1978; SANTOS et al. 1994).

In addition to inducing immune responses in the GALT, some oral antigen is absorbed. Although most dietary antigens are degraded by the time they reach the

Antigen and T lymphocyte traffic in the gut associated lymphoid tissue

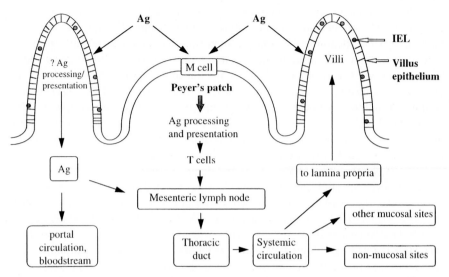

Fig. 1. Antigen and T lymphocyte traffic in the gut associated lymphoid tissue. Antigen is taken up via M cells into lymphoid nodules termed Peyer's patches or into the villus epithelium. Antigen processing and presentation results in the induction of cells which traffic to the systemic circulation via the mesenteric lymph node and thoracic duct and then migrate back to the lamina propria and to other mucosal and non-mucosal sites. Some antigen passes through the gut associated lymphoid tissue (GALT) into the bloodstream and portal circulation through or between epithelial cells. The villi contains intraepithelial lymphocytes which are activated CD8[1] T cells unique to the gut. T cells in the lamina propria are in a different state of activation than those in Peyer's patches. Particulate antigen may be preferentially taken up by M cells and soluble antigen by the villous epithelium. Peyer's patches also contain B cell rich poorly formed germinal centers where induction of antibody responses occur. *Ag*, antigen; *IEL*, intraepithelial lymphocyte

small intestine, studies in humans and rodents have demonstrated that some intact or partially degraded antigen is absorbed into the systemic circulation (BRUCE and FERGUSON 1986a; HUSBY et al. 1986; PENG et al. 1990). This absorbed or processed antigen may also be involved in inducing tolerance, although the mechanism associated with this pathway has not been defined.

3 Cytokine Milieu

It is known that the lymphoid tissue microenvironment can determine the outcome of antigen presentation to T cells. Indeed, the cytokine milieu present in the GALT strongly influences the generation of T cells which provide help to induce and differentiate IgA + B cells (MACDONALD 1983). It also favors the generation of regulatory T cells (after oral administration of protein) which can suppress antigen-specific immunological events in the periphery (CHEN et al. 1994; MILLER et al. 1992b; SANTOS et al. 1994).

The predominant type of immunoglobulin secreted by the B lymphocytes of the gut is IgA, accounting for 70%–90% of all Ig secreted in normal intestinal mucosa. IgA production in the GALT is important to trap potential antigens in the lumen before they elicit an immune response (ABREU-MARTIN and TARGAN 1996). The generation of secretory IgA in the gut is the combined result of T helper (Th) cytokines responsible for the switch from IgM to IgA (CEBRA et al. 1991). Transforming growth factor (TGF-β), derived from both T cells and nonlymphoid cells, and interleukin-5 (IL-5), derived from Th2 cells, favor the switch from IgM to IgA in mucosal B lymphocytes, while terminal differentiation of soluble IgA (sIgA)-positive B cells to IgA-secreting plasma cells is predominantly regulated by IL-5 and IL-6 (COFFMAN et al. 1989; DEEM and TARGAN 1995).

High levels of IL-4 and IL-5 mRNA were detected in LPLs and mesenteric lymph nodes after activation with phorbol myristate acetate (PMA) and ionomycin (JAMES et al. 1990). LPLs also demonstrated high expression of IL-2 and interferon-γ (IFN-γ). Similar results were seen in IELs with the additional production of IL-6 and TGF-β. Of these subsets LPLs had the highest number of cells spontaneously secreting IL-5. In the PP there is a high level of TGF-β production and relatively few IFN-γ- and IL-5-secreting cells compared to LPLs.

Although both Th1 and Th2-type cytokines are produced by the GALT, the overall effect is Th2 predominantly and achieves IgA rearrangement and secretion (ABREU-MARTIN and TARGAN 1996; JAMES et al. 1990). Furthermore, T cells in lymphoid organs drained by mucosal sites secrete IL-4 as a primary T cell growth factor, whereas those drained by nonmucosal sites secrete IL-2 (DAYNES et al. 1990) and there is induction of IL-4, IL-10 and TGF-β within 6 h of feeding TCR transgenic mice (GONNELLA et al. 1998).

These results indicate the uniqueness of the GALT since it favors the induction of Th2 cells and T cells that secrete TGF-β.

4 Mechanisms of Oral Tolerance

The primary mechanisms by which oral tolerance is mediated include deletion, anergy and active cellular suppression; the determining factor in this process is the dose of antigen fed (FRIEDMAN and WEINER 1994; WEINER et al. 1994). Low doses favor active suppression, whereas high doses favor deletion and anergy (Fig. 2). Nonetheless, there is evidence showing that these mechanisms may not be mutually exclusive (CHEN et al. 1997, 1995a). This review will focus on the generation of regulatory T cells in the gut upon oral administration of antigen and the ability of these cells to suppress inflammatory events in the periphery.

The exact site of tolerance induction for any given antigen has not been clearly established. The candidates are PP, the epithelial layer or the systemic lymphoid

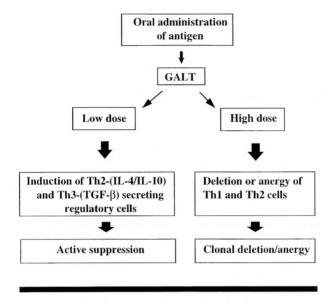

ORAL TOLERANCE

Fig. 2. Mechanisms of oral tolerance. *GALT*, gut associated lymphoid tissue; *IL*, interleukin; *TGF-β*, transforming growth factor β

compartment (exposed to antigenic fragments generated by digestion in the intestine) (BRUCE and FERGUSON 1986b; PENG et al. 1990). Of these possibilities, it appears that the PP is one of the main sites for the generation of regulatory cells (MILLER et al. 1992b; SANTOS et al. 1994).

It has been shown that, after oral administration of sheep red blood cells, ovalbumin or myelin basic protein (MBP), T cells are generated in PP which migrate to the spleen and lymph nodes generating peripheral tolerance (ANDRE et al. 1975; BITAR and WHITACRE 1988; HIGGINS and WEINER 1988; MATTINGLY and WAKSMAN 1978; RICHMAN et al. 1978). Recent evidence indicates that one of the primary mechanisms of active cellular suppression is via the secretion of suppressive cytokines such as TGF-β, IL-4 and IL-10 following antigen specific triggering (CHEN et al. 1994, 1996; MARON et al. 1996; MILLER et al. 1992b, 1993).

Stimulation of PP T cells from MBP fed animals with MBP in vitro resulted in secretion of TGF-β, a potent immunosuppressive cytokine (MILLER et al. 1992b; SANTOS et al. 1994). Indeed, cells from PP of animals fed MBP adoptively transferred protection to experimental autoimmune encephalomyelitis (EAE). T cells generated in the gut after MBP feeding in rats were CD8 + while in mice they were CD4 +. Both phenotypes secrete TGF-β and are able to protect EAE by adoptive transfer (LIDER et al. 1989; MILLER et al. 1992b).

For a time it was thought that immunological tolerance was achieved by deletion of autoreactive cells in the thymus but it is now clear that autoreactive cells can be found at about the same frequency in normal and autoimmune patients (HAFLER and WEINER 1995). Why these cells become activated and cause disease in some individuals, whereas in others they remain harmless, is a major question in basic immunology. How to control the autoimmune process once it has been initiated is the major problem in clinical medicine. Extensive investigation in both human and animals models of autoimmunity (BIEGANOWSKA et al. 1997; FUKAURA et al. 1996; HAFLER et al. 1988; ZHANG et al. 1993) indicate that it is the activation state and secretion of specific cytokines by autoreactive T cells that leads to self-tissue destruction and disease.

Thus, the induction of oral tolerance to self-antigens has become a important tool to manipulate the immune system toward a state of inhibition of the inflammatory events that lead to autoimmunity. It has been shown in several animal models of autoimmune diseases that antigen-specific regulatory T cells generated in the GALT migrate to lymphoid organs and suppress immune responses by inhibiting the generation of effector cells (WEINER 1997). They also traffic to the target organ where they suppress disease by releasing anti-inflammatory cytokines such as IL-4, IL-10 and TGF-β (KHOURY et al. 1992).

TGF-β-secreting $CD4^+$ cells specific for MBP have been cloned from the mesenteric lymph nodes of SJL mice (CHEN et al. 1994). These clones were found to be structurally identical to Th1 disease-inducing clones with respect to T cell receptor (TCR) usage, major histocompatibility complex (MHC) restriction and epitope recognition, but suppressed rather than induced disease. TGF-β-secreting $CD4^+$ cells were also cloned from mice transgenic for an MBP-specific TCR by culturing in the presence of IL-4, but not IL-2; these clones did not secrete IL-2, IFN-γ, IL-4 or IL-10, and used IL-4 as a growth/differentiation factor (INOBE et al. 1998). Seder et al. have also recently found that IL-4, and in some instances IL-10 and TGF-β, may serve to help differentiation of TGF-β secreting cells (SEDER et al. 1998). Thus, $CD4^+$ cells that primarily produce TGF-β appear to be a unique T cell subset that includes mucosal T helper function and down-regulatory properties for Th1 and other immune cells. In contrast to Th1 and Th2 cells, these cells provide help for IgA production and have been termed Th3 cells (CHEN et al. 1994; MOSMANN and SAD 1996; WEINER 1997) (Table 1). MBP- and proteolipid protein (PLP)-specific TGF-β-secreting Th3 type cells have been observed in multiple sclerosis patients orally administered myelin proteins (FUKAURA et al. 1996). Th3 clones grow poorly and characterization of these cells in terms of their relationship to Th1 and Th2 lineages has been slow. The fact that Th3 cells do not proliferate well may also account for their relative resistance to deletion, as compared with Th1 and Th2 cells, following feeding of high dose antigen to ovalbumin (OVA)-TCR-transgenic mice (CHEN et al. 1995a). Th3 type cells appear distinct from the Th2 cells as $CD4^+$ TGF-β-secreting cells with suppressive properties in the gut have been generated from IL-4-deficient animals (POWRIE et al. 1996). In addition, administration of anti-IL-12 to OVA-TCR transgenic animals is associated with inhibition of Th1 responses and high TGF-β production without increased IL-4 production (MARTH et al. 1996). As well as clones that only secrete TGF-β, other clones have been isolated that have mixed cytokine secretion patterns; these clones

Table 1. Characteristics of T cell subsets

	Th1	Th2	Th3	Tr1
Cytokine profile				
IFN-γ	+ + + +	–	+/–	+
IL-4	–	+ + + +	+/–	–
TGF-β	+/–	+/–	+ + + +	+ +
IL-10	–	+ +	+/–	+ + + +
Growth/differentiation factors	IL-2	IL2/IL4	IL-4/TGF-β	IL-10
Functions				
Help	DTH/IgG2a	IgG1/IgE	IgA	?
Suppression	Th2	Th1	Th1/Th2	Th1

DTH, delayed-type hypersensitivity; IFN-γ, interferon γ; IL, interleukin; TGF-β, transforming growth factor β; Th, T helper.

secrete IL-4 or IFN-γ, as well as TGF-β, and have suppressive properties (Inobe et al. 1998). Recently Bridoux, et al (1997) described TGF-β-dependent inhibition of Th2-induced autoimmunity by self MHC class II-specific regulatory CD4+ T cell lines. These TGF-β producing autoreactive T cells were distinct from classical Th1 or Th2 cells and were found to inhibit both Th1- and Th2-mediated autoimmune disease. Another type of regulatory cell is driven by IL-10 and secretes primarily IL-10 in addition to TGF-β and has been termed a Tr1 cell (Groux et al. 1997).

5 Bystander Suppression and Oral Tolerance

Bystander suppression was discovered during investigation of regulatory cells induced by oral administration of low doses of MBP (Miller et al. 1991). It addressed a major conceptual problem related to designing antigen- or T cell-specific therapy of inflammatory autoimmune diseases, such as multiple sclerosis , type I diabetes, and rheumatoid arthritis in which the autoantigen is unknown or where there are reactivities to multiple autoantigens in the target tissue.

In animal models of autoimmunity, during the course of chronic inflammatory autoimmune process, there is an intra- and interantigenic spread of autoreactivity at the target organ (Cross et al. 1993; Lehmann et al. 1992). Similar findings have been observed in human autoimmune disease in which there are reactivities to multiple autoantigens from the target tissue (Zhang et al. 1993). For example, in multiple sclerosis there is immune reactivity to three myelin antigens, MBP, PLP, and myelin-oligodendrocyte glycoprotein (MOG). In type I diabetes, there are multiple islet cell antigens that could be targets of autoreactivity, including glutamate decarboxylate, insulin, and heat shock proteins (Harrison 1992). Because regulatory cells induced by oral antigen secrete antigen-nonspecific cytokines after being triggered by the fed antigen, they suppress inflammation in the microenvironment where the antigen is localized. Thus, for an organ-specific inflammatory disease, one need not know the specific antigen that is the target of an autoimmune response but only feed an antigen capable of inducing regulatory cells that then migrate to the target tissue and suppress inflammation (WEINER 1997).

Bystander suppression was demonstrated in vitro when it was shown that cells from MBP-fed animals suppressed proliferation of an ovalbumin line across a transwell, but only when triggered by MBP. The soluble factor mediating this effect was identified as TGF-β (MILLER et al. 1992b; SANTOS et al. 1994). Bystander suppression has also been demonstrated in autoimmune disease models (AL-SAB-BAGH et al. 1996; YOSHINO et al. 1995) (Table 2). One can suppress PLP peptide-induced EAE by feeding MBP, and MBP-specific T cell clones from orally tolerized animals that secrete TGF-β also suppress PLP-induced disease (AL-SABBAGH et al. 1994). Other examples include the suppression of adjuvant- and antigen-induced arthritis by feeding type II collagen, the suppression of insulitis in the non-obese diabetic (NOD) mouse by feeding glucagon, and the suppression of lymphocytic choriomeningitis virus (LCMV)-induced diabetes in mice that have LCMV proteins expressed under the control of the insulin promoter on the pancreatic islet by feeding insulin (VON HERRATH et al. 1996; WEINER et al. 1994; YOSHINO et al. 1995; ZHANG and MICHAEL 1990).

In theory, bystander suppression could be applied to the treatment of organ-specific inflammatory conditions that are not classic autoimmune diseases, such as psoriasis, or could be used to target anti-inflammatory cytokines to an organ where inflammation may play a role in disease pathogenesis even if the disease is not primarily inflammatory in nature. For example, oral MBP decreased stroke size in a rat stroke model, presumably by decreasing inflammation associated with is-chemic injury (BECKER et al. 1997). Although bystander suppression was initially described in association with regulatory cells induced by oral antigen, the process could in principle be induced by any immune manipulation that induces Th2- or Th3-type regulatory cells.

Table 2. Models of autoimmune and other diseases that demonstrate bystander suppression

Disease	Immunizing antigen	Fed antigen	Target organ
EAE	PLP	MBP	Brain
EAE	MBP peptide 71–90	MBP peptide 21–40	Brain
EAE	MBP	OVA	Lymph node DTH responses
Arthritis	BSA, mycobacteria	Type II collagen	Joint
Diabetes	LCMV	Insulin	Pancreatic islets
Stroke	–	MBP	Brain

BSA, bovine serum albumin; DTH, delayed-type sensitivity; HSA, heat stable antigen; LCMV, lymphocytic choriomenigitis virus; MBP, myelin basic protein; OVA, ovalbumin; PLP, proteolipid protein.

6 Modulation of Oral Tolerance

If oral tolerance is broadly defined as the inhibition of Th1 responses in the periphery, either by large doses of antigen that induce anergy and/or deletion or by gut-induced Th2- or Th3-type regulatory cells, anything that favors Th1 vs Th2 or Th3 responses would abrogate oral tolerance and the converse (Table 3). Thus, large doses of IFN-γ abrogate oral tolerance (ZHANG and MICHAEL 1990) and similar results would be expected with IL-12 (MARINARO et al. 1997). Indeed investigators have found that anti-IL-12 enhances oral tolerance in animals transgenic for an OVA-specific TCR and is associated both with increased TGF-β secretion and T cell apoptosis (MARTH et al. 1996). In the uveitis model, IL-2 potentiates oral tolerance and is associated with increased production of TGF-β, IL-4 and IL-10 in PP (RIZZO et al. 1994). Furthermore, IL-4 administration enhances low dose oral tolerance to MBP in the EAE model and is associated with increased fecal IgA anti-MBP antibodies (INOBE et al. 1998). Lipopolysaccharide (LPS) enhances oral tolerance to MBP (KHOURY et al. 1990) and is associated with increased expression of IL-4 in the brain (KHOURY et al. 1992), and IFN-β synergizes with the induction of oral tolerance in SJL/J mice fed low doses of MBP (NELSON et al. 1996). Given its antagonistic effects on IL-12, IL-10 might be expected to enhance oral tolerance.

Cholera toxin (CT) is one of the most potent mucosal adjuvants and feeding CT abrogates oral tolerance when fed with an unrelated protein antigen (ELSON and EALDING 1984). However, when a protein is coupled to recombinant CT B subunit (CTB) and given orally, there is enhancement of peripheral immune tolerance (BERGEROT et al. 1997; SUN et al. 1994, 1996). Although it is not known how CTB enhances oral tolerance, it may do so by the more efficient induction of regulatory cells due to the special binding properties of the CTB subunit. The abrogation of oral tolerance by CT may also be the result of an active immune process in the gut, as it has been shown that CT given orally with keyhole lympet hemocyanin (KLH) primes both Th1 and Th2 responses (HÖRNQUIST and LYCKE 1993), even though Th2 responses may be preferentially generated (WILSON et al. 1991; XU-AMANO

Table 3. Modulation of oral tolerance

Augments	Decreases
IL-2	IFN-γ
IL-4/IL-10[a]	IL-12
Anti-IL-12	CT
CTB antigen	Anti-MCP-1
LPS	Anti-γδ-A6
IFN-β	GVH
Multiple emulsions	

CTB, cholera toxin B subunit; IFN, interferon; IL, interleukin; LPS, lipopolysaccharide; MCP, monocyte chemotactic protein 1.
[a] H.L. Weiner and A. Slavin, unpublished.

et al. 1990). The priming of Th1 responses via the gut may makes it impossible to suppress systemic Th1 responses following peripheral immunization. It has also been reported that antibodies to the chemokine monocyte chemotactic protein 1 (MCP-1) abrogate oral tolerance, perhaps by interfering with chemotaxis of the Th2- or Th3-type regulatory cells generated in the gut or by acting as a factor required for generation of such cells (KARPUS and LUKACS 1996). Oral antigen delivery using a multiple emulsion system also enhances oral tolerance (ELSON et al. 1996). Orally administered cytokines retain their biologic activity (ROLLWAGON and BAQAR 1996) and oral IL-4 and IL-10 can enhance oral tolerance (INOBE et al. 1998).

Further insight into mechanisms of oral tolerance and the factors that modulate it are beginning to emerge with the investigation of gene-knockout or -deficient animals. Although impaired mucosal immune responses have been reported in IL-4-deficient animals (VAJDY et al. 1995), induction of oral tolerance was possible using high doses of antigen (GARSIDE et al. 1995b). Experiments are now required to determine if oral tolerance in these animals is secondary to regulatory cell generation or anergy and/or deletion. Some investigators were unable to induce oral tolerance in IFN-γ-deficient animals (KJERRULF et al. 1996) whereas others have reported induction of oral tolerance in animals deficient in the IFN-γ receptor (IFN-γR) (KWEON et al. 1996). These differences may relate to the antigen doses and the use of cholera toxin. γδT cells appear to play an important role in oral tolerance although the mechanism of action is poorly understood. Animals deficient in γδT cells have impaired mucosal IgA responses (FUJIHASHI et al. 1996) and it has been reported to be more difficult to induce oral tolerance in animals depleted of such cells (KE and KAPP 1996; KE et al. 1997) (Weiner and Spahn, unpublished). As discussed above, oral tolerance can also be induced in CD8-deficient animals (CHEN et al. 1995b; GARSIDE et al. 1995a; TADA et al. 1996). A lack of oral tolerance and priming for contact sensitivity was reported in MHC class II-deficient mice and CD4[+] T cell-depleted mice (DESVINGES et al. 1996). Anti-B7-2 antibody blocks induction of low dose, but not high dose, oral tolerance (H.L. Weiner and L.M. Liu, unpublished). Commonly used drugs for autoimmune diseases may affect oral tolerance; for instance, methotrexate, a drug used to treat rheumatoid arthritis, enhances oral tolerance (AL-SABBAGH et al. 1997).

7 Manipulation of Th Responses in Experimental Models of Autoimmunity

Data from several experimental autoimmune models suggest that cytokines associated with the Th1 phenotype of lymphocytes, i.e., IFN-γ, TNF-β, tumor necrosis factor-α (TNF-α), and IL-12, promote inflammation while cytokines associated with the Th2 subset, IL-4, IL-10 and TGF-β, have a role in suppressing disease (WEINER 1997; WEINER and MAYER 1996). Of note is the observation that natural recovery from EAE is associated with down-regulation of inflammatory cytokines and differential up-regulation of TGF-β, IL-4 and prostaglandin E (PGE) expression in the brain (KHOURY et al. 1992). However, the complexity and redundancy of the immune system can be well demonstrated by the induction of EAE in IFN-γ knockout mice (FERBER et al. 1996) and more recently in TNF knockout mice (FREI et al. 1997).

The finding that Th1 and Th2 subsets of CD4+ T cell can cross-regulate each other via production of distinct cytokine patterns supports the idea that oral tolerance induced by low doses of antigen reflects the down-regulation of Th1 CD4+ cells by Th2 cells. This is further demonstrated by a large series of studies from several laboratories that have demonstrated that orally administered autoantigens can induce the generation of Th2/Th3 cells which down-regulate Th1 cells suppressing several experimental models of autoimmunity (Table 4).

8 Experimental Autoimmune Encephalomyelitis

Detailed immunohistology was performed in animals orally tolerized with MBP and in animals naturally recovering from EAE (KHOURY et al. 1992). Brains from OVA fed control animals at the peak of disease showed perivascular infiltration with activated mononuclear cells which secreted the inflammatory cytokines IL-1, IL-2 TNF-α, IFN-γ, IL-6 and IL-8. Inhibitory cytokines TGF-β and IL-4 and PGE2 were absent. In contrast, animals orally tolerized with MBP showed a marked reduction of the perivascular infiltrate and down-regulation of all inflammatory cytokines. In addition, there was up-regulation of the inhibitory cytokine TGF-β.

Table 4. Suppression of autoimmunity by oral tolerance

Animals models	Model	Protein fed
	EAE	MBP, PLP
	Arthritis (CII, AA)	CII
	Uveitis S	Ag, IRBP
	Myasthenia gravis	AChR
	Diabetes (NOD mouse)	Insulin, GAD
	Transplantation	Alloantigen, MHC peptide
	Thyroiditis	Thyroglobulin
	Colitis	Haptenized colonic proteins
Human disease trials	*Disease trial*	*Protein fed*
	Multiple sclerosis	Bovine myelin
	Rheumatoid arthritis	Chicken type II collagen
	Uveitis	Bovine S-Ag
	Type 1 diabetes	Human insulin

AChR, acetylcholine receptor; CII, type II collagen; EAE, experimental allergic encephalomyelitis; IRBP, interphotoreceptor binding protein; MBP, myelin basic protein; MHC, major histocompatibility complex; NOD, nonobese diabetic; PLP, proteolipid protein.

It was further shown that mucosally derived Th3-type clones induced by oral antigen can actively regulate immune responses in vivo (CHEN et al. 1994). T cell clones isolated from the mesenteric lymph nodes of SJL mice that had been orally tolerized to MBP produced TGF-β, with various amounts of IL-4 and IL-10, and suppressed EAE induced with either MBP or PLP.

Furthermore, it was demonstrated that MBP-specific T cells can differentiate in vivo into encephalitogenic or regulatory T cells depending upon the context by which they are exposed to antigen (CHEN et al. 1996). Transgenic mice for a TCR derived from an encephalitogenic T cell clone specific for the peptide of MBP, Ac-1–11, plus I-Au restricted were fed low or high doses of MBP, and without further immunization spleen cells were tested for cytokine production. Low dose feeding induced prominent secretion of IL-4, IL-10 and TGF-β whereas minimal secretion of these cytokines was observed with high dose feeding. Adoptive transfer of these cells markedly suppressed the induction of EAE. In contrast to oral tolerization, subcutaneous immunization of transgenic mice with MBP in complete Freund's adjuvant induced IFN-γ secreting Th1 cells in vitro and experimental encephalo-myelitis in vivo. These data support the idea that the same T cell could attain the ability to become an encephalitogenic or a regulatory T cell depending upon the context in which it is exposed to antigen. In addition, this differentiation is dose-dependent when oral antigen is administered.

9 Insulin-Dependent Diabetes Mellitus

The NOD mouse spontaneously develops an insulin-dependent diabetes mellitus (IDDM) that has many immunological and pathological similarities to human type I diabetes. Th1 cells that secrete IL-2, IFN-γ and TNF-α and support cell-mediated immunity have been implicated in the pathogenesis of diabetes (LIBLAU et al. 1995). Similarly, Th2 cells can have a protective role in IDDM.

Pancreas tissues of NOD mice fed OVA or insulin were analyzed for cytokine production and cell types (HANCOCK et al. 1995; MARON et al. 1996). At 10 weeks of age, control NOD mice or mice fed with OVA showed dense infiltration of mononuclear cells (MNCs) consisting mainly of CD4+ T cells within pancreas tissues. The MNCs were associated with dense expression of IFN-γ and TNF-α, as well as focal IL-2 and IL-2 receptor (IL-2R) expression, but lacked labeling for IL-4, IL-10, TGF-β, or PGE. In contrast to control NOD mice, oral insulin admin-istration markedly altered the extent of insulitis and cytokine expression. Oral insulin not only decreased the overall incidence of insulitis but suppressed to background levels the extent of IL-2R expression and eliminated production of IL-2, IFN-γ, or TNF-α. NOD mice fed insulin also showed labeling for IL-4, IL-10, TGF-β, and PGE.

T cells from mice fed insulin B-chain produced less IFN-γ than control mice and a moderate level of TGF-β. In addition, T cell lines generated from NOD mice orally tolerized to insulin B-chain produced high amounts of IL-4 and IL-10 and did not produce IFN-γ and IL-2. Adoptive transfer of this insulin B-chain Th2 cell line suppressed the onset of diabetes in NOD mice and delayed the onset of disease when it was cotransferred with diabetic cells.

10 Collagen-Induced Arthritis

One of the first studies to demonstrate that an orally administered autoantigen can suppress an autoimmune disease was the use of type II collagen in collagen-induced arthritis (CIA) (NAGLER-ANDERSON et al. 1986; THOMPSON and STAINES 1986).

Recent analysis of cytokine expression in rheumatoid arthritis tissues and in experimental models of rheumatoid arthritis have provided a better understanding of the actual role of each cytokine in the context of the inflammatory process. Thus, it was shown that many proinflammatory cytokines such as TNF-α, IL-1, IL-6, granulocyte/macrophage colony-stimulating factor (GM-CSF) and chemokines such as IL-8 are abundant in rheumatoid arthritis tissues (FELDMAN et al. 1996). This inflammatory milieu is compensated to some degree by the increased production of anti-inflammatory cytokines such as IL-10 and TGF-β and cytokine inhibitors such as IL-1 receptor antagonist (IL-1ra) and soluble TNF receptor (TNF-R). However, these regulatory mechanisms are not sufficient to prevent tissue damage.

TGF-β has been of particular interest because of its vast range of actions – from inflammatory to immunosuppressive effects (WAHL 1994). The destruction of cartilage is now considered to be mostly due to the activity of matrix metalloproteinases (MMPs), produced by activated macrophages and fibroblasts in response to IL-1 and TNF-α. TGF-β not only inhibits the production of these proinflammatory cytokines but also induces the production of their native inhibitors, known as tissue inhibitors of metalloproteinase (TIMPs) (WRIGHT et al. 1991). TGF-β is also involved in tissue repair as it is able to stimulate the production of type I and type XI collagen (KHALIL et al. 1989). Thus, locally expressed TGF-β may promote a reparative process in arthritic synovial tissue by inhibiting cartilage and bone destruction. However, in chronic lesions, overproduction of TGF-β could contribute to the ongoing damage by recruiting inflammatory cells with the potential for tissue destruction. It has been shown that injection of TGF-β locally in the joints of normal rats results in a rapid leukocyte infiltration with synovial hyperplasia (FAVA et al. 1991). Interestingly, whereas systematically administered TGF-β attenuates joint inflammation in arthritic rats (WAHL 1992) and mice (THORBECKE et al. 1992), consistent with its known immunosuppressive activities, local (intraarticular) injection surprisingly exacerbated the inflammatory process in the joints (HINES et al. 1993; WAHL 1992; WAHL et al. 1993).

IL-10 is well known to inhibit IFN-γ production by Th1 cells but can also inhibit the production of a variety of cytokines from other leukocyte populations

(Moore et al. 1993). It inhibits the production of IL-1, IL-6, TNF-α, IL-8, G-CSF from macrophages and IL-1, TNF, IL-8, macrophage inflammatory protein (MIP)-1α and MIP-1β from polymorphonuclear cells. Most of these cytokines and chemokines have been implicated in the pathological processes of arthritis (Kasama et al. 1995).

We have investigated the effects of oral administration of type II collagen (CII) on the cytokine pattern and proliferative response of T cells following footpad immunization with the same antigen. T cells from mice fed collagen showed suppression of proliferative responses and IFN-γ production compared to control mice (OVA fed). Furthermore, T cell lines derived from orally tolerized mice produced more IL-4 and IL-10 and less IFN-γ than T cell lines from control mice (Table 5). Similar observations were made after nasal administration of collagen. It was also shown that inhibition of CIA by oral and nasal administration of collagen was correlated with a suppression of proliferation and IFN-γ production.

Thus, our results show that oral and nasal administration of low doses of type II collagen can inhibit the generation and activation of antigen-specific Th1 cells. Furthermore, long-term cell lines producing anti-inflammatory cytokines can be generated from mice treated by oral and nasal route with CII. These anti-inflammatory cells might be involved in the suppression of CIA saw upon oral and nasal treatment with CII. It is interesting to note that mice treated by either oral and nasal administration of CII that became sick showed diminished proliferative responses in vitro and did not produce IFN-γ compared to control. This immunological status can be related to the mild outcome of disease seen in these mice.

Several studies have demonstrated the ability of IL-4 and IL-10 to suppress the inflammatory process and tissue damage in experimental models of arthritis (Chomarat et al. 1995; Miossec et al. 1992). Intraperitoneal administration of recombinant IL-4 inhibited an experimental model of arthritis by decreasing the influx of inflammatory cells to the tissue, decreasing TNF-α production and reactive oxygen intermediate metabolism. It also increased the production of IL-1ra (Allen et al. 1993). This protein can modulate inflammation and tissue degrada-

Table 5. Proliferation and cytokine pattern of popliteal lymph node (PLN) cells and T cell lines from DBA/1 mice fed collagen type II before immunization with the same antigen

		Proliferation (Δ CPM)	IFN-γ pg/ml	IL-10 pg/ml	IL-4 pg/ml	TGF-β Δpg/ml
PLN	CII fed	4964	0	0	0	650
	PBS fed	15109	1650	0	0	70
T cell lines	CII fed	–	1750	1400	640	0
	PBS fed	–	15600	500	0	0

DBA/1 mice were fed 30 µg/day of collagen type II (CII) or PBS for 5 consecutive days. Two days after the last feeding animals were immunized in the footpad with 100 µg of CII + complete Freund's adjuvant (CFA). Ten days later PLN cells were stimulated in vitro with antigen and proliferative response and cytokine production in the supernatants were measured. T cell lines derived from the two groups (CII fed and PBS fed) were cultured in vitro and restimulated with CII and antigen-presenting cells (APCs) every 2 weeks. The result shown above is from fifth stimulation.

tion by blocking IL-1-induced PGE2 and collagenase and neutralizing IL-1 induction of IL-6.

In contrast to TFG-β and IL-10, IL-4 has not been found in rheumatoid arthritis synovial tissue cultures nor in T cells cloned from rheumatoid arthritis synovial biopsies (FELDMAN et al. 1996).The generation of T cells secreting IL-4 and IL-10 by oral administration of antigen may therefore be a useful therapeutic agent.

11 Treatment of Autoimmune Diseases in Humans

Exposure of a contact-sensitizing agent via the mucosa prior to subsequent skin challenge led to unresponsiveness in a portion of patients studied (LOWNEY 1968). KLH administered orally to human subjects has been reported to decrease subsequent cell-mediated immune responses although antibody responses were not affected (HUSBY et al. 1994). Nasal KLH has also been reported to induce tolerance in humans (WALDO et al. 1994). On the basis of the long history of oral tolerance and the safety of the approach, human trials have been initiated in multiple sclerosis, rheumatoid arthritis, uveitis and diabetes (Table 4). These initial trials suggest that there has been no systemic toxicity or exacerbation of disease. Although positive effects have been observed, consistent clinical efficacy has yet to be demonstrated. Results in humans, however, have paralleled several aspects of what has been observed in animals.

In multiple sclerosis patients, MBP- and PLP-specific TGF-β-secreting Th3-type cells have been observed in the peripheral blood of patients treated with an oral bovine myelin preparation and not in patients who were untreated (FUKAURA et al. 1996). There was no increase in MBP- or PLP-specific IFN-γ-secreting cells in treated patients. These results demonstrate that it is possible to immunize via the gut for autoantigen-specific TGF-β-secreting cells in a human autoimmune disease by oral administration of the autoantigen. However, a recently completed 515 patient, placebo-controlled, double-blind Phase III trial of single-dose bovine myelin in relapsing-remitting multiple sclerosis did not show differences between placebo and treated groups in the number relapses, a large placebo effect being observed (AutoImmune, Inc., Lexington, MA, USA). Preliminary analysis of magnetic resonance imaging data showed significant changes favoring oral myelin in certain patient subgroups.

In rheumatoid arthritis, a single center double-blind study of oral collagen showed differences favoring placebo (TRENTHAM et al. 1993). A recently completed 280 patient, double-blind Phase II dosing trial of chicken type II collagen in doses ranging from 20µg to 2500µg demonstrated statistically significant positive effects in the group treated with the lowest dose (BARNETT et al. 1998). Oral administration of larger doses of bovine type II collagen (1–10 mg) did not show a significant difference between tested and placebo groups, although a higher prevalence of responders was reported for the groups treated with type II collagen (SIEPER et al. 1996). These results are consistent with animal studies of orally administered type II collagen, in which protection against adjuvant- and antigen-induced arthritis and bystander suppression was observed only at the lower doses (YOSHINO et al. 1995; ZHANG et al. 1990). An open-label pilot study of oral collagen in juvenile rheumatoid arthritis gave positive results with no toxicity (BARNETT et al. 1996). This lack of systemic toxicity is an important feature for the clinical use of oral tolerance, especially in children for whom the long-term effects of immunosuppressive drugs is unknown. Five Phase II randomized studies of oral type II collagen have been performed and, based on the results obtained, a multicenter double-blind Phase III trial study of oral type II collagen (Colloral) is underway (Autoimmune Inc.). The five double-blind Phase II studies involved a total of 805 patients treated with oral type II collagen and 296 treated with placebo. Two of the studies have been published (TRENTHAM 1993; BARNETT 1998). The other three studies were included in an integrated analysis that led to the decision to carry out a Phase II trial. A dose refinement study tested doses of 5, 20, and 60μg. An abrupt methotrexate withdrawal study tested the effect of 20μg oral type II collagen in patients abruptly withdrawn from methotrexate, and a hydroxychloroquine comparison trial compared 20μg oral type II collagen with hydroxychloroquine. Analysis of study interactions suggested that integrating clinical data was appropriate to gain information to decide upon a Phase III trial. Using linear logistic regression, a statistically significant effect favoring patients treated with oral type II collagen vs those on placebo was found, as measured by obtaining a Paulus 20 response at any time during the study (cumulative Paulus). The effectiveness of individual Colloral doses was tested, checking cumulative Paulus 20/50 and 70 criteria. Colloral at 60μg was found to be the most significant dose, compared to other doses. Safety analysis demonstrated that Colloral was extraordinarily safe, with no side effects, and appears to be the safest of the current medicines given for rheumatoid arthritis. The magnitude of the clinical responses to Colloral appear to be on the same level as nonsteroidal anti-inflammatory drugs (NSAIDs) for the majority of patients. However, there is a subgroup of patients that appear to have a more significant response to the medication. The studies to date are encouraging for the potential use of oral collagen in the treatment of rheumatoid arthritis in the future.

In uveitis, a pilot trial of S antigen (S-Ag) and an S-Ag mixture has recently been completed at the National Eye Institute (Bethesda, MD, USA) and showed positive trends with oral bovine S-Ag but not the retinal mixture (NUSSENBLATT et al. 1997). Trials have been initiated in new-onset diabetes in which recombinant human insulin is administered orally, and trials are underway in subjects at risk for diabetes as part of the diabetes prevention trial (DPT-1). Preliminary analysis of a randomized, double-blind, placebo con trolled study of oral insulin in newly diagnosed immune-mediated (type I) diabetes demonstrated preserved β-cell function, as measured by endogenous C-peptide insulin responses in adult new-onset diabetics fed 10 mg recombinant human insulin as compared to those fed placebo (COUTANT et al. 1998). Oral desensitization to nickel allergy in humans induces a decrease in nickel-specific T cells and affects cutaneous eczema (BAGOT et al. 1995). Finally, a pilot immunological study of oral MHC peptides has been initiated in transplantation patients. Based on results to date in humans, it appears that the clinical administration of oral antigen for the treatment of human conditions will depend on the specific disease and the nature and dosages of proteins administered, and in some instances may require the use of synergists or mucosal adjuvants to enhance biologic effects. Also, recombinant human proteins may be more efficacious than animals proteins (MILLER et al. 1992a).

12 Future Directions

Although it is clear that oral antigen can suppress autoimmunity in animals, much remains to be learned. Under certain experimental conditions worsening of autoimmune diseases in animals by oral antigen has been reported (BLANAS et al. 1996; MEYER et al. 1996; MILLER et al. 1994; TERATO et al. 1996) although this has not been observed in human trials. Cell surface molecules and cytokines associated with inductive events in the gut that generate and modulate oral tolerance are not completely understood. Important areas of investigation include cytokine milieu, antigen presentation and costimulation requirements, routes of antigen processing, form of the antigen, role of the liver, the effect or oral antigens on antibody and IgE responses and on cytotoxic T lymphocytes, and the role of $\gamma\delta T$ cells. As the molecular events associated with the generation and modulation of oral tolerance are better understood, the ability to apply mucosal tolerance successfully to the treatment of human autoimmune and other diseases will be further enhanced.

References

Abreu-Martin MT, Targan SR (1996) Lamina propria lymphocytes: A unique population of mucosal lymphocytes. In: Kagnoff MF, H Kiyono (eds) Essentials of mucosal immunology Academic, San Diego, pp 227–245

Al-Sabbagh A, Garcia G, Slavin A, Weiner HL, Nelson P (1997) Combination therapy with oral myelin basic protein and oral methotrexate enhances suppression of experimental autoimmune encephalomyelitis. Neurology 48:A421

Al-Sabbagh A, Miller A, Santos LMB, Weiner HL (1994) Antigen-driven tissue-specific suppression following oral tolerance: orally administered myelin basic protein suppresses proteolipid induced experimental autoimmune encephalomyelitis in the SJL mouse. Eur J Immunol 24:2104–2109

Al-Sabbagh A, Nelson P, Sobel RA, Weiner HL (1996) Antigen-driven peripheral immune tolerance: suppression of experimental autoimmune encephalomyelitis and collagen induced arthritis by aerosol administration of myelin basic protein or type II collagen. Cell Immunol 171:111–119

Allen JB, Wong HL, Costa GL, Bienkowski MJ, Wahl SM (1993) Suppression of monocyte function and differential regulation of IL-1 and IL-1ra by IL-4 contribute to resolution of experimental arthritis. J Immunol 151:4344–4351

Andre C, Heremans JF, Vaerman JP, Cambiaso CL (1975) A mechanism for the induction of immunological tolerance by antigen feeding: antigen-antibody complexes. J Exp Med 142:1509–1519

Bagot M, Charue D, Flechet ML, Terki N, Toma A, Revuz J (1995) Oral desensitization in nickel allergy induces a decrease in nickel-specific T-cells. Eur J Dermatol 5:614–617

Barnett ML, Kremer JM, St. Claair EW, Clegg DO, Furst D, Weisman M, Fletcher MJF, Lavin PT, Finger E, Morales A, Le CH, Trentham DE (1998) Treatment of rheumatoid arthritis with oral type II collagen: results of a multicenter double-blind placebo controlled trial. Arthritis Rheum 41:290–297s

Barnett ML, Combitchi D, Trentham DE (1996) A pilot trial of oral type II collagen in the treatment of juvenile rheumatoid arthritis. Arthritis Rheum 39:623–628

Becker KJ, McCarron RM, Hallenbeck JM (1997) Oral tolerance to myelin basic protein decreases stroke size. Stroke 28:246

Bergerot I, Fioix C, Peterson J, Moulin V, Rask C, Fabien N, Lindblad M, Mayer A, Czerkinsky C, Holmgren J, Thivolet C (1997) A cholera toxoid-insulin conjugate as an oral vaccine against spontaneous autoimmune diabetes. Proc Natl Acad Sci USA 94:4610–4614

Bieganowska KD, Ausubel LJ, Modabber Y, Slovik E, Messersmith W, Hafler DA (1997) Direct ex vico analysis of activated Fas-sensitive autoreactive T cells in human autoimmune disease. J Exp Med 185:1581–1594

Bitar D, Whitacre CC (1988) Suppression of experimental autoimmune encephalomyelitis by the oral administration of myelin basic protein. Cell Immunol 112:364–370

Blanas E, Carbone FR, Allison J, Miller JFAP, Heath WR (1996) Induction of autoimmune diabetes by oral administration of autoantigen. Science 274:1707–1709

Bridoux F, Badou A, Saoudi A, Bernard I, Druet E, Pasquier R, Druet P, Pelletier L (1997) Transforming growth factor b (TGF-b)-dependent inhibition of T helper cell 2 (Th2)-induced autoimmunity by self-major histocompatibility complex (MHC) class II-specific regulatory CD4+ T cell line. J Exp Med 185:1769–1775

Bruce MG, Ferguson A (1986a) The influence of intestinal processing on the immunogenicity and molecular size of absorbed circulating ovalbumin in mice. Immunology 59:295–300

Bruce MG, Ferguson A (1986b) Oral tolerance to ovalbumin in mice: studies of chemically modified and of "biologically filtered" antigen. Immunology 57:627–30

Cebra JJ, George A, Schrader CE (1991) A microculture containing TH2 and dendritic cells supports the production of IgA by clones from both primary and IgA memory B cells and by single germinal center B cells from Peyer's patches. Immunol Res 10:389–392

Challacombe SJ, Tomasi TB (1980) Systemic tolerance and secretory immunity after oral immunization. J Exp Med 152:1459–1472

Chase M (1946) Inhibition of experimental drug allergy by prior feeding of the sensitizing agent. Proc Soc Exp Biol Med 61:257–259

Chen Y, Hancock W, Marks R, Gonnella PA, Weiner HL (1997) Mechanisms of recovery from cell-mediated autoimmune disease: T cell deletion and immune deviation following encephalomyelitis in myelin basic protein T cell receptor transgenic mice. Cell Immunol (in press)

Chen Y, Inobe J-I, Kuchroo VK, Baron JL, Janeway CA, Weiner HL (1996) Oral tolerance in myelin basic protein T-cell receptor transgenic mice: suppression of autoimmune encephalomyelitis and dose-dependent induction of regulatory cells. Proc Natl Acad Sci USA 93:388–391

Chen Y, Inobe J-I, Marks R, Gonnella P, Kuchroo VK, Weiner HL (1995a) Peripheral deletion of antigen-reactive T cells in oral tolerance. Nature 376:177–180

Chen Y, Inobe J-I, Weiner HL (1995b) Induction of oral tolerance to myelin basic protein in CD-8 depleted mice: both CD4+ and CD8+ T cells mediate active suppression. J Immunol 156:910–916

Chen Y, Kuchroo VK, Inobe J-I, Hafler DA, Weiner HL (1994) Regulatory T cell clones induced by oral tolerance: suppression of autoimmune encephalomyelitis. Science 265:1237–1240

Chomarat P, Vannier E, Dechanet J, Dinnarelo CA, Miossec P (1995) Balance of III receptor antagonism/III beta in rheumatoid synovium and its regulation by IL-4 and IL-10. J Immunol 154:1432–39

Coffman R, Lebman D, Shrader B (1989) Transforming growth factor beta (TGFb) specifically enhances IgA production by lipopolysaccharide-stimulated murine B lymphocytes. J Exp Med 144:3411–16

Coutant R, Zeidler A, Rappaport R, Schatz D, Schwartz S, Raskin P, Rogers D, Bode B, Crockett S, Marks J, Deeb L, Chalew S, MacLauren N (1998) Oral insulin therapy in newly diagnosed immune mediated (type 1) diabetes. Preliminary analysis of a randomized, double-blind, placebo controlled study. Diabetes 47:A97

Cross AH, Tuohy VK, Raine CS (1993) Development of reactivity to new myelin antigens during chronic relapsing autoimmune demyelination. Cell Immunol 146:261–270

Daynes R, Araneo B, Dowell T, Huang K, Dudley D (1990) Regulation of murine lymphokine pro-
    duction in vivo III The lymphoid tissue microenvironment exerts regulatory influences over T helper
    cell function. J Exp Med 171:979 996
Deem RL, Targan SR (1995) The role of TGFb in molding the activation state of lamina propria T cells.
    Gastroenterology 108 (abstract)
Desvinges C, Bour H, Nicolas JF, Kaiserlian D (1996) Lack of oral tolerance but oral priming for contact
    sensitivity to dinitrofluorobenzene in major histocompatibility complex class II-deficient mice and in
    CD4 + T cell-depleted mice. Eur J Immunol 26:1756 1761
Elson CO, Ealding W (1984) Cholera toxin feeding did not induce oral tolerance in mice and abrogated
    oral tolerance to an unrelated protein antigen. J Immunol 133:2892 2897
Elson CO, Tomasi M, Dertzbaugh MT, Thaggard G, Hunter R, Weaver C (1996) Oral antigen delivery
    by way of a multiple emulsion system enhances oral tolerance. In: Weiner HL, Mayer LF (eds) Oral
    tolerance: mechanisms and applications. Ann New York Acad Sci, New York, pp 156 162
Fava RA, Olsen NJ, Postlethwaite AE, Broadley KN, Davidson JM, Naney LB, Lucas C, Townes AS
    (1991) TGFb induced neutrophil recruitment to synovial tissues. J Exp Med 173:1121 32
Feldman M, Brennan FM, Maini RN (1996) Role of cytokines in rheumatoid arthritis. Ann Rev Im-
    munol 14:397 440
Ferber IA, Brocke S, Taylor-Edwards C, Ridgway W, Dinisco C, Steinman L, Dalton D, Fathman CG
    (1996) Mice with disrupted IFN-g gene are susceptible to the induction of experimental autoimmune
    encephalomyelitis (EAE). J Immunol 156:5 7
Frei K, Eugster H-P, Bopst M, Constantinescu CS, Lavi E, Fontana A (1997) Tumour necrosis factor α
    and lymphotoxin α are not required for induction of acute experimental autoimmune en-
    cephalomyelitis. J Exp Med 185:2177 2182
Friedman A, Weiner HL (1994) Induction of eanergy or active suppression following oral tolerance is
    determined by antigen dosage. Proc Natl Acad Sci USA 91:6688 6692
Fujihashi K, McGhee JR, Kweon MN, Cooper MD, Tonegawa S, Takahashi I, Hiroi T, Mestecky
    J, Kiyono H (1996) g/d T Cell-deficient mice have impaired mucosal immunoglobulin A responses.
    J Exp Med 183:1929 1935
Fukaura H, Kent SC, Pietrusewicz MJ, Khoury SJ, Weiner HL, Hafler DA (1996) Induction of circu-
    lating myelin basic protein and proteolipid protein specific TGF-b1 secreting Th3 T cells by oral
    administration of myelin in multiple sclerosis patients. J Clin Invest 98:70 77
Garside P, Steel M, Liew FY, Mowat AM (1995a) CD4 + but not CD8 + T cells are required for the
    induction of oral tolerance. Int Immunol 7:501 504
Garside P, Steel M, Worthey EA, Satoskar A, Alexander J, Bluethmann H, Liew FY, Mowat M (1995b)
    T helper 2 cells are subject to high dose oral tolerance and are not essential for its induction.
    J Immunol 154:5649 5655
Gonnella PA, Chen Y, Inobe J-I, Quartulli M, Weiner HL (1998) In situ immune response in gut
    associated lymphoid tissue (GALT) following oral antigen in TcR transgenic mice. J Immunol
    160:4708 4718
Groux H, O'Garra A, Bigler M, Rouleau M, Antonenko S, de Vries JE, Roncarolo MG (1997) A CD4 +
    T cell subset inhibits antigen-specific T cell responses and prevents colitis. Nature 389:737 742
Hafler DA, Duby AD, Lee SJ, Benjamin D, Seidman JG, Weiner HL (1988) Oligoclonal T-lymphocyte in the
    cerebrospinal fluid of patients with inflammatory central nervous system diseases. J Exp Med 167:1313
Hafler DA, Weiner HL (1995) Immunologic mechanisms and therapy in multiple sclerosis. Immunol Rev
    144:75 107
Hancock WW, Polanski M, Zhang ZJ, Blogg N, Weiner HL (1995) Suppression of insulitis in NOD mice
    by oral insulin administration is associated with selective expression of IL-4 IL-10 TGF-b and pro-
    staglandin-E. Amer J Pathol 147:1193 1199
Harrison LC (1992) Islet cell antigens in insulin-dependent diabetes: Pandora's box revisited. Immunol
    Today 13:348 352
Higgins P, Weiner HL (1988) Suppression of experimental autoimmune encephalomyelitis by oral ad-
    ministration of myelin basic protein and its fragments. J Immunol 140:440 445
Hines KL, Christ M, Wahl M (1993) Cytokine regulation of the immune response: an in vivo model.
    Immunomethods 3:13 22
Hörnquist E , Lycke N (1993) Cholera toxin adjuvant greatly promotes antigen priming of T cells. Eur J
    Immunol 23:2136 2143
Husby S, Jensenius JC, Svehag S-E (1986) Passage of undegraded dietary antigen into the blood of
    healthy adults. Further characterization of the kinetics of uptake and the size distribution of the
    antigen. Scand J Immunol 24:447 452

Husby S, Mestecky J, Moldoveanu Z, Holland S, Elson CO (1994) Oral tolerance in humans: T cell but not B cell tolerance after antigen feeding. J Immunol 152:4663 4670

Inobe J-I, Slavin AJ, Komagata Y, Chen Y, Liu L, Weiner HL (1998) Il-4 is a differentiation factor for TGF-β secreting Th3 cells and oral administration of Il-4 enhances oral tolerance in experimental allergic encephalomyelitis. Eur J Immunol 28:2780 2790

James SP, Kwan WC, Sneller MC (1990) T cells in inductive and effector compartments of intestinal mucosal immune system of non-human primates differ in lymphokine mRNA expression lymphokine utilization and regulatory function. J Immunol 144:1251 1256

Kagnoff MF (1996) Essential of mucosal immunology. Kagnoff MF, Kagnoff KH (eds) Academic, San Diego

Karpus WJ, Lukacs NW (1996) The role of chemokines in oral tolerance: abrogation of nonrespon-siveness by treatment with antimonocyte chemotactic protein-1. In: Weiner HL, Mayer LF (eds) Oral tolerance: mechanisms and applications. Ann New York Acad Sci, New York, pp 133 144

Kasama T, Strieter RM, Lukacs NW, Lincoln PM, Burdick MD, Kunkel SL (1995) Il-10 expression and chemokine regulation during the evolution of murine type II CIA. J Clin Invest 95:2868 2876

Ke Y, Kapp JA (1996) Oral antigen inhibits priming of CD8+ CTL CD4+ T cells and antibody responses while activating CD8+ suppressor. T cells J Immunol 156:916 921

Ke Y, Pearce K, Lake JP, Ziegler HK, Kapp JA (1997) gd T lymphocytes regulate the induction and maintenance of oral tolerance. J Immunol 158:3610 3618

Khalil N, Bereznay O, Sporn M, Greenberg AH (1989) Macrophage production of TGFb and fibroblast collagen synthesis in the chronic pulmonary inflammation. J Exp Med 170:727 37

Khoury SJ, Hancock WW, Weiner HL (1992) Oral tolerance to myelin basic protein and natural recovery from experimental autoimmune encephalomyelitis is associated with downregulation of inflamma-tory cytokines and differential upregulation of transforming growth factor b interleukin 4 and pro-staglandin E expression in the brain. J Exp Med 176:1355 1364

Khoury SJ, Lider O, Al-Sabbagh A, Weiner HL (1990) Suppression of experimental autoimmune en-cephalomyelitis by oral administration of myelin basic protein III. Synergistic effect of lipopolysac-charide. Cell Immunol 131:302 310

Kjerrulf M, Grdic D, Lycke N (1996) Impaired mucosal IgA responses but intact oral tolerance in IFN-g receptor deficient mice. FASEB J 10:A1418

Kweon M, Fujihashi K, Yamamoto M, Van Cott JL, McGhee JR, Kiyono H (1996) Mucosally induced tolerance in IL-4−/−but not IFN-g −/−mice. FASEB J 10:A1028

Lehmann P, Forsthuber T, Miller A, Sercarz E (1992) Spreading of T-cell autoimmunity to cryptic determinants of an autoantigen. Nature 358:155 157

Liblau RS, Singer SM, McDevitt HO (1995) Th1 and Th2 CD4+ T cells in the pathogenesis of organ-specific autoimmune diseases. Immunol Today 16:34 38

Lider O, Santos LMB, Lee CSY, Higgins PJ, Weiner HL (1989) Suppression of experimental allergic encephalomyelitis by oral administration of myelin basic protein II. Suppression of disease and in vitro immune responses is mediated by antigen-specific CD8+ T lymphocytes. J Immunol 142:748 752

Lowney ED (1968) Immunologic unresponsiveness to a contact sensitizer in man. J Invest Dermatol 51:411 417

MacDonald TT (1983) Immunosuppression caused by antigen feeding II Suppressor T cells mask Peyer's patches B cell priming to orally administered antigen. Eur J Immunol 13:138 142

Marinaro M, Boyaka PN, Finkelman FD, Kiyono H, Jackson RJ, Jirillo E, McGhee JR (1997) Oral but not parental interleukin (IL)-12 redirects T helper 2 (Th2)-type responses to an oral vaccine without altering mucosal IgA responses. J Exp Med 185:415 427

Maron R, Blogg NS, Polanski M, Hancock W, Weiner HL (1996) Oral tolerance to insulin and the insulin B-chain. Ann N Y Acad Sci 778:346 357

Marth T, Strober W, Kelsall BL (1996) High dose oral tolerance in ovalbumin TCR-transgenic mice: systemic neutralization of IL-12 augments TGF-beta secretion and T cell apoptosis. J Immunol 157:2348 2357

Mattingly JA, Waksman BH (1978) Immunologic suppression after oral administration of antigen I. Specific suppressor cells formed in rat Peyer's patches after oral administration of sheep erythrocytes and their systemic migration. J Immunol 121:1878 1882

Meyer AL, Benson JM, Gienapp IE , Cox KL, Whitacre CC (1996) Suppression of murine chronic relapsing experimental autoimmune encephalomyelitis by the oral administration of myelin basic protein. J Immunol 157:4230 4238

Miller A, Al-Sabbagh A, Santos L, Das MP, Weiner HL (1993) Epitopes of myelin basic protein that trigger TGF-b release following oral tolerization are distinct from encephalitogenic epitopes and mediate epitope driven bystander suppression. J Immunol 151:7307–7315

Miller A, Lider O, Abramsky O, Weiner HL (1994) Orally administered myelin basic protein in neonates primes for immune responses and enhances experimental autoimmune encephalomyelitis in adult animals. Eur J Immunol 24:1026–1032

Miller A, Lider O, Al-Sabbagh A, Weiner H (1992a) Suppression of experimental autoimmune encephalomyelitis by oral administration of myelin basic protein V. Hierarchy of suppression by myelin basic protein from different species. J Neuroimmunol 39:243–250

Miller A, Lider O, Roberts AB, Sporn MB, Weiner HL (1992b) Suppressor T cells generated by oral tolerization to myelin basic protein suppress both in vitro and in vivo immune responses by the release of TGFb following antigen specific triggering. Proc Natl Acad Sci USA 89:421–425

Miller A, Lider O, Weiner HL (1991) Antigen-driven bystander suppression following oral administration of antigens. J Exp Med 174:791–798

Miller S, Hanson D (1979) Inhibition of specific immune responses by feeding protein antigens IV. Evidence for tolerance and specific active suppression of cell-mediated immune responses to ovalbumin. J Immunol 123:2344

Miossec P, Briolay J, Dechanet J, Wijdenes J, Martinez-Valdez H, Banchereau J (1992) Inhibition of the production of proinflammatory cytokines and Ig by IL-4 in an ex vivo model of rheumatoid synovitis. Arthritis Rheum 35:874–883

Moore KW, O'Garra A, de Waal Malefyt R, Viera P, Mosman TR (1993) Interleukin 10. Ann Rev Immunol 11:165–190

Mosmann TR, Sad S (1996) The expanding universe of T-cell subsets: Th1 Th2 and more. Immunol Today 17:138–146

Mowat AM (1987) The regulation of immune responses to dietary protein antigens. Immunol Today 8:93–98

Nagler-Anderson C, Bober LA, Robinson ME, Siskind GW, Thorbeke FJ (1986) Suppression of type II collagen-induced arthritis by intragastric administration of soluble type II collagen. Proc Natl Acad Sci USA 83:7443–7446

Nelson PA, Akselband Y, Dearborn S, Al-Sabbagh A, Tian ZJ, Gonnella P, Zamvil S, Chen Y, Weiner HL (1996) Effect of oral beta interferon on subsequent immune responsiveness. In: Weiner HL, Mayer LF (eds) Oral tolerance: mechanisms and applications. New York Acad Sci, New York, pp 145–155

Neutra RM, Kraehenbuhl J-P (1996) M cells as a pathway for antigen uptake and processing. In: Kagnoff MF, Kiyono H (eds) Essentials of mucosal immunology Academic, San Diego, pp 29–33

Newby TJ (1984) Protective immune responses in the intestinal tract In: Newby TJ, Stokes CR (eds) Local immune responses of the gut. CRC, Boca Raton, Florida, pp 143–198

Ngan J, Kind LS (1978) Suppressor T cells for IgE and IgG in Peyer's patches of mice made tolerant by the oral administration of ovalbumin. J Immunol 120:861–865

Nussenblatt RB, Gery I, Weiner HL, Ferris F, Shiloach J, Ramaley N, Perry C, Caspi R, Hafler DA, Foster S, Whitcup SM (1997) Treatment of uveitis by oral administration of retinal antigens: Results of a Phase I/II randomized masked trial. Am J Opthalmol 123:583–592

Peng HJ, Turner MW, Strobel S (1990) The generation of a "tolerogen" after the ingestion of ovalbumin is time-dependent and unrelated to serum levels of immunoreactive antigen. Clin Exp Immunol 81:510–515

Powrie F, Carlino J, Leach MW, Mauze S, Coffman RL (1996) A critical role for transforming growth factor-beta but not interleukin 4 in the suppression of T helper type 1-mediated colitis by CD45RB (low) CD4+ T cells. J Exp Med 183:2669–2674

Richman LK, Chiller JM, Brown WR, Hanson DG, Vaz NM (1978) Enterically induced immunological tolerance I Induction of suppressor T lymphocytes by intragastric administration of soluble proteins. J Immunol 121:2429–2433

Rizzo LV, Miller-Rivero NE, Chan C-C, Wiggert B, Nussenblatt RB, Caspi RR (1994) Interleukin-2 treatment potentiates induction of oral tolerance in a murine model of autoimmunity. J Clin Invest 94:1668–1672

Rollwagon FM and Baqar S (1996) Oral cytokine administration. Immunol Today 17:548–550

Santos LMB, Al-Sabbagh A, Londono A, Weiner HL (1994) Oral tolerance to myelin basic protein induces regulatory TGF-β-secreting T cells in Peyer's patches of SJL mice. Cell Immunol 157:439–447

Seder RA, Marth T, Sieve MC, Strober W, Letterio JJ, Roberts AB, Kelsall B (1998) Factors involved in the differentiation of TGF-β producing cells from naive CD4+ T cells: IL-4 and IFN-γ have opposing effects, while TGF-β positively regulates its own production. J Immunol 160:5719–5728

Sieper J, Kary S, Sörensen H, Alten R, Eggens U, Hüge W, Hiepe F, Kühne A, Listing J, Ulbrich N, Braun J, Zink A, Mitchison NA (1996) Oral type II collagen treatment in early rheumatoid arthritis. Arthritis Rheum 39:41–51

Strobel S (1990) Mechanisms of gastrointestinal immunoregulation and food induced injury to the gut. Eur J Clin Nutr 45:1–9

Sun J-B, Holmgren C, Czerkinsky C (1994) Cholera toxin B subunit: an efficient transmucosal carrier-delivery system for induction of peripheral immunological tolerance. Proc Natl Acad Sci USA 91:10795–10799

Sun JB, Rask C, Olsson T, Holmgren J, Czerkinsky C (1996) Treatment of experimental autoimmune encephalomyelitis by feeding myelin basic protein conjugated to cholera toxin B subunit. Proc Natl Acad Sci USA 93:7196–7201

Tada Y, Ho A, Koh DR, Mak TW (1996) Collagen-induced arthritis in CD4- or CD8- deficient mice. J Immunol 156:4520–4526

Terato K, Xiu JY, Miyahara H, Cremer MA, Griffiths MM (1996) Induction of chronic autoimmune arthritis in DBA/1 mice by oral administration of type II collagen and Escherichia coli lipopoly-saccharide. Brit J Rheumatol 35:1–11

Thompson HS, Staines NA (1986) Suppression of collagen-induced arthritis with pregastrically or in-travenously administered type II collagen. Agents Actions 19:318–319

Thorbecke GJ, Shah R, Leu CH, Kuruvilla AP, Hardison AM, Palladino MA (1992) Involvement of endogenous TNFa and TGFb during induction of collagen type II arthritis in mice. Proc Natl Acad Sci USA 89:7375–7379

Trentham DE, Dynesius-Trentham RA, Orav EJ, Combitchi D, Lorenzo C, Sewell KL, Hafler DA, Weiner HL (1993) Effects of oral administration of type II collagen on rheumatoid arthritis. Science 261:1727–1730

Vajdy M, Kosco-Vilbois MH, Kopf M, Köhler G, Lycke N (1995) Impaired mucosal immune responses in interleukin 4-targeted mice. J Exp Med 181:41–53

Vaz NM, Maia LCS, Hanson DG, Lynch JM (1977) Inhibition of homocytotropic antibody responses in adult inbred mice by previous feeding of the specific antigen. J Allergy Clin Immunol 60:110–115

von Herrath MG, Dyrberg T, Oldstone MBA (1996) Oral insulin treatment suppresses virus-induced antigen-specific destruction of beta cells and prevents autoimmune diabetes in transgenic mice. J Clin Invest 98:1324–1331

Wahl S (1992) Transforming growth factor beta (TGFb) in inflammation: a cause and a cure. J Clin Immunol 12:61–74

Wahl SM (1994) TGFb: the good the bad and the ugly. J Exp Med 180:1587–90

Wahl SM, Allen JB, Costa GL, Wong HL, Dasch JR (1993) Reversal of acute and chronic synovial inflammation by anti-TGFb. J Exp Med 177:225–230

Waldo FB, Van Den Wall Bake AWL, Mestecky J, Husby S (1994) Suppression of the immune response by nasal immunization. Clin Immunol Immunopathol 72:30–34

Weiner HL (1997) Oral tolerance: immune mechanisms and treatment of autoimmune diseases. Immunol Today 18:335–343

Weiner HL, Friedman A, Miller A, Khoury SJ, Al-Sabbagh A, Santos LMB, Sayegh M, Nussenblatt RB, Trentham DE, Hafler DA (1994) Oral tolerance: Immunologic mechanisms and treatment of animal and human organ-specific autoimmune diseases by oral administration of autoantigens. Ann Rev Immunol 12:809–837

Weiner HL, Inobe J-I, Kuchroo V, Chen Y (1996) Induction and Characterization of TGF-beta secreting Th3 cells. FASEB J 10:A1444

Weiner HL, Mayer LF (1996) Oral tolerance: mechanisms and applications. Ann N Y Acad Sci, New York, pp 1–453

Wilson AD, Bailey M, Williams NA, Stokes CR (1991) The in vitro production of cytokines by mucosal lymphocytes immunized by oral administration of keyhole limpet hemocyanin using cholera toxin as an adjuvant. Eur J Immunol 21:2333–2339

Wright JK, Cawston TE, Hazelman BL (1991) TGFb stimulates the production of tissue inhibitor of metalloproteinases (TIMP) by human synovial and skin fibroblasts. Biochem Biophys Acta 1094:207–210

Xu-Amano J, Kiyono H, Jackson RJ, Staats HF, Fujihashi K, Burrows PD, Elson CO, Pillai S, McGhee JR (1990) Helper T cell subsets for immunoglobulin A responses: oral immunization with tetanus toxoid and cholera toxin as adjuvant selectively induces Th2 cells in mucosa associated tissues. J Exp Med 178:1309–1320

Yoshino S, Quattrocchi E, Weiner HL (1995) Oral administration of type II collagen suppresses antigen-induced arthritis in Lewis rats. Arthritis Rheum 38:1092 1096

Zhang J, Markovic S, Raus J, Lacet B, Weiner HL, Hafler DA (1993) Increased frequency of IL-2 responsive T cells specific for myelin basic protein and proteolipid protein in peripheral blood and cerebrospinal fluid of patients with multiple sclerosis. J Exp Med 179:973 984

Zhang JZ, Lee CSY, Lider O, Weiner HL (1990) Suppression of adjuvant arthritis in Lewis rats by oral administration of type II collagen. J Immunol 145:2489 2493

Zhang Z, Michael J (1990) Orally inducible immune unresponsiveness is abrogated by IFN-g treatment. J Immunol 144:4163 4165

Subject Index

A
abortion, unexplained recurrent   38
allergen-specific immunotherapy
   39
allergic diseases   1
altered peptide ligands   42
anergy   40, 126
anti-IFN-γ 92
anti-IL-4   90
anti-IL-12   94
anti-TGF-β1   92
anti-TNF   93
anti-TNF/LT   93
antigen therapy   96
apoptosis   97
autoimmune disease   79 – 122
– therapies   84, 137 – 139
autoimmunity
– antigen-specific therapies   89
– development   83

B
bystander   129

C
CCR3   29
CCR4   29
CD4$^+$ NK1.1$^+$ T cells   31
CD30   29
cholera toxin   131
class II MHC molecules   110
collagen-induced arthritis   82, 135
costimulatory molecules   109
CpG dinucleotides   45
CXCR4   29
cytokines   84 – 87
– in autoimmune diseases   90 – 94

D
D. farinae group 1 allergen   43
deletion   126
diabetes   81
DNA, naked   44

D
DO11.10 TCR system   15
DTH   2

E
experimental autoimmune encephalitis   80, 96,
   100, 106

F
FcεR$^+$ non-T cells   30

G
gastrointestinal tract   123
GATA-3   33
genetic dysregulation   35, 36
gut-associated lymphoid tissue (GALT)   124

H
hepatitis A   38
humanized anti-human IgE antibodies   46
hygiene hypothesis   38

I
IELs (intraepithelial lymphocytes)   124
IFN-α 6, 16, 34, 42
IFN-γ 14, 16, 34, 42, 92
– signaling   20
– producing Th1 cells   2
IFN-γR-/-   92
IgE   2
– antibodies, humanized anti-human   46
IL-2   6
IL-4   16, 90
– antagonizing   46
– producing Th2 cells   2
– Th cell-derived   35
IL-5   46
IL-10   3, 60, 90, 91
IL-12   3, 14, 33, 42, 60 – 72, 94
– as adjuvant for infectious diseases   64 – 67
– antagonists as therapy   87 – 89
– in vitro activities   62
– immunotherapeutic agent   69, 70
– induction of Th1 responses   63

IL-12
– role in infections   64
– β1 subunit   16
– β2 subunit   17
– molecular control of production   61
– in tumor therapy   67–69
IL-12-/-   94
IL-13   7, 91
insulin dependent diabetes mellitus   81
– specific antigen therapy   102–105
intraepithelial lymphocytes (IELs)   124

J
JAK2   59

L
Leishmania
– *brasiliensis*   43
– *major*   2, 42
leprosy   3
– lepromatous   3
– tuberculoid   3
lymphotoxin   92

M
Mycobacterium
– *tuberculosis*   38
– *bovis*   43
measles infection   38
microorganisms, recombinant   43
multiple sclerosis   80

N
neonatal tolerance   111
NOD   134

O
oral tolerance   123
– modulation   131

P
peptides   40
Peyer's patches   124
polyI: C   42

R
recombinant microorganisms   43
rheumatoid arthritis   82

S
Salmonella   43
specific antigen therapy   105, 106
STAT   15, 16
STAT4 (transcription factor)   59
STAT6   32
sTNFR   93
sTNFR-Fc   93
suppression, active   126
surface markers   29

T
T cells
– IFN-γ producing   57
– naive   31
γδ T cells   132
targeting Th2 cells   45
targeting transcription factors   46
TGF-β   127
– secreting CD4⁺ cells   128
TGF-β1   91
TGF-β2   91
Th responses   123–139
Th1 cells
– definition   29
– development   14, 20
– IFN-γ-producing   2
– inducing cytokines   42
– mediated diseases   80
– stability and reversibility   1–10
Th2 cells   28, 29
– allergen-specific   40
– – redirection of allergen-specific responses
    41–45
– definition   29
– development   14, 20, 34, 36
– – environmental factors   37
– IL-4 producing   2
– stability and reversibility   1–10
– targeting   45
TNF   93
TYK2   59

Printing: Saladruck, Berlin
Binding: Buchbinderei Lüderitz & Bauer, Berlin

Current Topics in Microbiology and Immunology

Volumes published since 1989 (and still available)

Vol. 198: **Griffiths, Gillian M.; Tschopp, Jürg (Eds.):** Pathways for Cytolysis. 1995. 45 figs. IX, 224 pp. ISBN 3-540-58725-X

Vol. 199/I: **Doerfler, Walter; Böhm, Petra (Eds.):** The Molecular Repertoire of Adenoviruses I. 1995. 51 figs. XIII, 280 pp. ISBN 3-540-58828-0

Vol. 199/II: **Doerfler, Walter; Böhm, Petra (Eds.):** The Molecular Repertoire of Adenoviruses II. 1995. 36 figs. XIII, 278 pp. ISBN 3-540-58829-9

Vol. 199/III: **Doerfler, Walter; Böhm, Petra (Eds.):** The Molecular Repertoire of Adenoviruses III. 1995. 51 figs. XIII, 310 pp. ISBN 3-540-58987-2

Vol. 200: **Kroemer, Guido; Martinez-A., Carlos (Eds.):** Apoptosis in Immunology. 1995. 14 figs. XI, 242 pp. ISBN 3-540-58756-X

Vol. 201: **Kosco-Vilbois, Marie H. (Ed.):** An Antigen Depository of the Immune System: Follicular Dendritic Cells. 1995. 39 figs. IX, 209 pp. ISBN 3-540-59013-7

Vol. 202: **Oldstone, Michael B. A.; Vitković, Ljubiša (Eds.):** HIV and Dementia. 1995. 40 figs. XIII, 279 pp. ISBN 3-540-59117-6

Vol. 203: **Sarnow, Peter (Ed.):** Cap-Independent Translation. 1995. 31 figs. XI, 183 pp. ISBN 3-540-59121-4

Vol. 204: **Saedler, Heinz; Gierl, Alfons (Eds.):** Transposable Elements. 1995. 42 figs. IX, 234 pp. ISBN 3-540-59342-X

Vol. 205: **Littman, Dan R. (Ed.):** The CD4 Molecule. 1995. 29 figs. XIII, 182 pp. ISBN 3-540-59344-6

Vol. 206: **Chisari, Francis V.; Oldstone, Michael B. A. (Eds.):** Transgenic Models of Human Viral and Immunological Disease. 1995. 53 figs. XI, 345 pp. ISBN 3-540-59341-1

Vol. 207: **Prusiner, Stanley B. (Ed.):** Prions Prions Prions. 1995. 42 figs. VII, 163 pp. ISBN 3-540-59343-8

Vol. 208: **Farnham, Peggy J. (Ed.):** Transcriptional Control of Cell Growth. 1995. 17 figs. IX, 141 pp. ISBN 3-540-60113-9

Vol. 209: **Miller, Virginia L. (Ed.):** Bacterial Invasiveness. 1996. 16 figs. IX, 115 pp. ISBN 3-540-60065-5

Vol. 210: **Potter, Michael; Rose, Noel R. (Eds.):** Immunology of Silicones. 1996. 136 figs. XX, 430 pp. ISBN 3-540-60272-0

Vol. 211: **Wolff, Linda; Perkins, Archibald S. (Eds.):** Molecular Aspects of Myeloid Stem Cell Development. 1996. 98 figs. XIV, 298 pp. ISBN 3-540-60414-6

Vol. 212: **Vainio, Olli; Imhof, Beat A. (Eds.):** Immunology and Developmental Biology of the Chicken. 1996. 43 figs. IX, 281 pp. ISBN 3-540-60585-1

Vol. 213/I: **Günthert, Ursula; Birchmeier, Walter (Eds.):** Attempts to Understand Metastasis Formation I. 1996. 35 figs. XV, 293 pp. ISBN 3-540-60680-7

Vol. 213/II: **Günthert, Ursula; Birchmeier, Walter (Eds.):** Attempts to Understand Metastasis Formation II. 1996. 33 figs. XV, 288 pp. ISBN 3-540-60681-5

Vol. 213/III: **Günthert, Ursula; Schlag, Peter M.; Birchmeier, Walter (Eds.):** Attempts to Understand Metastasis Formation III. 1996. 14 figs. XV, 262 pp. ISBN 3-540-60682-3

Vol. 214: **Kräusslich, Hans-Georg (Ed.):** Morphogenesis and Maturation of Retroviruses. 1996. 34 figs. XI, 344 pp. ISBN 3-540-60928-8

Vol. 215: **Shinnick, Thomas M. (Ed.):** Tuberculosis. 1996. 46 figs. XI, 307 pp. ISBN 3-540-60985-7

Vol. 216: **Rietschel, Ernst Th.; Wagner, Hermann (Eds.):** Pathology of Septic Shock. 1996. 34 figs. X, 321 pp. ISBN 3-540-61026-X

Vol. 217: **Jessberger, Rolf; Lieber, Michael R. (Eds.):** Molecular Analysis of DNA Rearrangements in the Immune System. 1996. 43 figs. IX, 224 pp. ISBN 3-540-61037-5

Vol. 218: **Berns, Kenneth I.; Giraud, Catherine (Eds.):** Adeno-Associated Virus (AAV) Vectors in Gene Therapy. 1996. 38 figs. IX,173 pp. ISBN 3-540-61076-6

Vol. 219: **Gross, Uwe (Ed.):** Toxoplasma gondii. 1996. 31 figs. XI, 274 pp. ISBN 3-540-61300-5

Vol. 220: **Rauscher, Frank J. III; Vogt, Peter K. (Eds.):** Chromosomal Translocations and Oncogenic Transcription Factors. 1997. 28 figs. XI, 166 pp. ISBN 3-540-61402-8

Vol. 221: **Kastan, Michael B. (Ed.):** Genetic Instability and Tumorigenesis. 1997. 12 figs.VII, 180 pp. ISBN 3-540-61518-0

Vol. 222: **Olding, Lars B. (Ed.):** Reproductive Immunology. 1997. 17 figs. XII, 219 pp. ISBN 3-540-61888-0

Vol. 223: **Tracy, S.; Chapman, N. M.; Mahy, B. W. J. (Eds.):** The Coxsackie B Viruses. 1997. 37 figs. VIII, 336 pp. ISBN 3-540-62390-6

Vol. 224: **Potter, Michael; Melchers, Fritz (Eds.):** C-Myc in B-Cell Neoplasia. 1997. 94 figs. XII, 291 pp. ISBN 3-540-62892-4

Vol. 225: **Vogt, Peter K.; Mahan, Michael J. (Eds.):** Bacterial Infection: Close Encounters at the Host Pathogen Interface. 1998. 15 figs. IX, 169 pp. ISBN 3-540-63260-3

Vol. 226: **Koprowski, Hilary; Weiner, David B. (Eds.):** DNA Vaccination/Genetic Vaccination. 1998. 31 figs. XVIII, 198 pp. ISBN 3-540-63392-8

Vol. 227: **Vogt, Peter K.; Reed, Steven I. (Eds.):** Cyclin Dependent Kinase (CDK) Inhibitors. 1998. 15 figs. XII, 169 pp. ISBN 3-540-63429-0

Vol. 228: **Pawson, Anthony I. (Ed.):** Protein Modules in Signal Transduction. 1998. 42 figs. IX, 368 pp. ISBN 3-540-63396-0

Vol. 229: **Kelsoe, Garnett; Flajnik, Martin (Eds.):** Somatic Diversification of Immune Responses. 1998. 38 figs. IX, 221 pp. ISBN 3-540-63608-0

Vol. 230: **Kärre, Klas; Colonna, Marco (Eds.):** Specificity, Function, and Development of NK Cells. 1998. 22 figs. IX, 248 pp. ISBN 3-540-63941-1

Vol. 231: **Holzmann, Bernhard; Wagner, Hermann (Eds.):** Leukocyte Integrins in the Immune System and Malignant Disease. 1998. 40 figs. XIII, 189 pp. ISBN 3-540-63609-9

Vol. 232: **Whitton, J. Lindsay (Ed.):** Antigen Presentation. 1998. 11 figs. IX, 244 pp. ISBN 3-540-63813-X

Vol. 233/I: **Tyler, Kenneth L.; Oldstone, Michael B. A. (Eds.):** Reoviruses I. 1998. 29 figs. XVIII, 223 pp. ISBN 3-540-63946-2

Vol. 233/II: **Tyler, Kenneth L.; Oldstone, Michael B. A. (Eds.):** Reoviruses II. 1998. 45 figs. XVI, 187 pp. ISBN 3-540-63947-0

Vol. 234: **Frankel, Arthur E. (Ed.):** Clinical Applications of Immunotoxins. 1999. 16 figs. IX, 122 pp. ISBN 3-540-64097-5

Vol. 235: **Klenk, Hans-Dieter (Ed.):** Marburg and Ebola Viruses. 1999. 34 figs. XI, 225 pp. ISBN 3-540-64729-5

Vol. 236: **Kraehenbuhl, Jean-Pierre; Neutra, Marian R. (Eds.):** Defense of Mucosal Surfaces: Pathogenesis, Immunity and Vaccines. 1999. 30 figs. IX, 296 pp. ISBN 3-540-64730-9

Vol. 237: **Claesson-Welsh, Lena (Ed.):** Vascular Growth Factors and Angiogenesis. 1999. 36 figs. X, 189 pp. ISBN 3-540-64731-7